COMMUNITY CAMPAIGNS FOR SUSTAINABLE LIVING: HEALTH, WASTE & PROTEST IN CIVIL SOCIETY

ADVANCES IN ECOPOLITICS

The Transition to Sustainable Living: Advances in Ecopolitics Volume 4
Edited by Liam Leonard and John Barry

Global Ecological Politics: Advances in Ecopolitics Volume 5
Edited by Liam Leonard and John Barry

Sustainable Justice and the Community Volume 6
By Liam Leonard and Paula Kenny

ADVANCES IN ECOPOLITICS VOLUME 7

COMMUNITY CAMPAIGNS FOR SUSTAINABLE LIVING: HEALTH, WASTE & PROTEST IN CIVIL SOCIETY

BY

LIAM LEONARD
School of Business and Humanities,
Institute of Technology, Sligo, Republic of Ireland

Emerald

United Kingdom – North America – Japan
India – Malaysia – China

Emerald Group Publishing Limited
Howard House, Wagon Lane, Bingley BD16 1WA, UK

First edition 2011

Reprints and permission service
Contact: booksandseries@emeraldinsight.com

British Library Cataloguing in Publication Data
A catalogue record for this book is available from the British Library

ISBN: 978-1-78052-380-4
ISSN: 2041-806X (Series)

CONTENTS

ABOUT THE AUTHOR

Dr. Liam Leonard is the series editor of the *Advances in Ecopolitics* Book Series with Emerald Group Publishing. Having worked for the National University of Ireland for a number of years as lecturer in Environmental Politics and Social Movements and Civil Society, he currently works as lecturer in Sociology, Criminology and Human Rights at the Institute of Technology, Sligo, Ireland. He is member of the Executive Committee of the Sociology Association of Ireland, and is the founder and chair of the Green Link West think tank, as well as founder and editor of the *Journal of Social Criminology*.

This book is dedicated to Shane and Caitríona Leonard.

PREFACE

The seventh edition in the *Advances in Ecopolitics* Series with Emerald Group Publishing is titled *Community Campaigns for Sustainable Living: Health, Waste & Protest in Civil Society*. This book is based on research and observations undertaken for my PhD thesis at the National University of Ireland between 2000 and 2004, and represents a case study of national and regional campaigns against both the Irish state's Regional Waste Management Plans and the corporate sector's attempts to develop waste incinerators or dumps in various parts of Ireland. For the most part, the study focuses on one western Irish city, Galway, a city often called 'The fastest growing city in Europe' during the boom years of the 1990s.

This book provides an in-depth account of the mobilising patterns and framing processes of community campaigns which emerged in the wake of the Irish state's introduction of regional plans for waste management, which included plans for municipal 'waste to energy' plants or incinerators. The book details a case study on community movements which emerged in response to waste plans by the state and corporate sector. It outlines these cases by examining the mobilisation patterns and framing processes of community social movements with health and environmental concerns.

It is the only book with a sole focus on this aspect of Irish society during the 'Celtic Tiger' boom which preceded the current economic downturn and examines policy, population, development social issues and local and national electoral processes in detail at a time of immense change in the Republic of Ireland. As such, it provides a salient insight into the societal shifts which provide opportunities for social movements to oppose state or corporate plans which may be perceived to have human health or environmental risks associated with them. However, while this book focuses on issues in Irish society, the environmental justice movement has its roots in the United States.

THE ANTI-TOXICS MOVEMENT IN THE UNITED STATES

Background

The anti-toxics movement's origins can be traced back to Rachel Carson's *A Silent Spring* in 1962. The book highlighted the impact of pesticides such as DDT on plant and wildlife in America in the years following the introduction of scientific methods of agriculture in the United States. In the aftermath of a wider public concern and scientific debate about Carson's work, President John F. Kennedy called on the Science Advisory Committee to investigate issues surrounding the use of pesticides. This inquiry confirmed Carson's position, and led to the regulation of the use of chemical pesticides in the United States. Carson has been subject to a number of subsequent criticisms from scientists working for the chemical industry.

The environmental justice movement developed further during the 1960s when migrant agricultural workers led by Cesar Chavez also challenged the use of pesticides in California, United States. African-American communities mobilised under a racial justice and anti-toxics agenda as part of a number of regional campaigns against toxic plants in Houston, Texas and Harlem, in New York City during the 1960s. In 1978, communities campaigned against the dumping of toxics near their homes in the Love Canal township near Niagara Falls, New York in the United States. The campaign emerged in response to concerns about high rates of cancer and birth defects in the area. The toxic waste had been buried in the region by the Hooker Chemical Company in the 1920s and had begun to seep into the local water supplies.

The land surrounding Love Canal was developed for a school and housing despite the warnings of the Hooker Company in the 1950s. Over 50 drums of chemical waste were found at the site during excavations, and the school was built away from the area. In 1957 low-income housing was built on the site. With the construction of a motorway in the 1970s, floods began to occur in the area, often containing toxic residue. In 1978, Lois Gibbs of 'the Love Canal Homeowners Association' led residents in protests about the number of serious illnesses occurring with children in the Love Canal community. The activities of the Homeowners Association led to the discovery of the toxic materials underneath the homes in the area. This information had been withheld from the community when the homes were first built. The Homeowners Association found that over 50% of their residents suffered in some way from the effects of the toxic waste underneath their housing.

US President Jimmy Carter allocated funds to assist with the Love Canal controversy, which had begun to make national headlines and found coverage in television news broadcasts. An investigation by the Environmental Protection Agency (EPA) in 1979 found that the Love Canal area suffered from an abnormally high number of serious illnesses and miscarriages. Pregnant women were evacuated from the Love Canal area as a precaution. The EPA Report also found that up to a third of residents had detectable damage to their chromosomes due to exposure to the chemicals at the site. Love Canal was declared a national emergency site in 1980. Over 700 families were evacuated and re-housed, and the Comprehensive Environmental Response, Compensation, and Liability Act (CERCLA) or 'Superfund' Act was signed into law, and in 1995 the EPA won nearly 300 million dollars in compensation for the incident. Legislation against toxic industries continued to be introduced in the aftermath of increased agitation from the public. Many industries would relocate to nations that lacked similar legislation, but would find that local opposition would occur as communities discovered the effects of toxics in emissions over time.

The community-based environmental justice movement developed further in the 1980s in response to the siting of hazardous plants or dumps, often in economically disadvantaged non-white neighbourhoods. This section of the anti-toxics movement developed in conjunction with the civil rights movement and was organised in response to repeated incidents of siting toxic plants or dumps in minority neighbourhoods. Suspicions about toxic industries were heightened by the industrial accidents at Three Mile Island, Chernobyl and Bhopal in India. Campaigns against incinerators emerged in the 1980s. The emissions from the incineration of waste released dioxins and furans into the atmosphere, and communities near incineration plants began to oppose the incineration industry as a result. While the concept of 'Waste to Energy' (WTE) was attractive to many municipalities due to the dualistic outcomes of waste incineration and energy creation from the same process, concerns raised about emissions by the anti-toxics movement were gradually extended to incineration plants.

Campaigns Against Incineration in the United States

The first major campaigns against incineration in the Unite States occurred at the same time as a 'garbage crisis' was being highlighted in the media. This crisis emerged as a result of regional landfills reaching capacity, and in some cases closing down. In addition, scientists such as Barry Commoner and Paul Connet were highlighting the carcinogenic risks posed by incinerator emissions. Another contributory factor was the discovery of a document prepared for the incineration industry in California which suggested ways of overcoming 'Political Difficulties Facing Waste to Energy conversion Plant Siting'. The document, sometimes known as 'the Cerrell Report' after the company that had commissioned it, was a handbook in overcoming community resistance to incinerators. It fell into anti-incineration activist's possession, and would later be used to demonstrate the technocratic approach taken by the industry when dealing with local communities.

Lois Gibbs of the Love Canal campaign had formed 'the Citizens Clearinghouse for Hazardous Wastes' (CCHW) in the early 1980s. The CCHW would prove to be a major resource for the anti-toxics and anti-incineration movements. The CCHW provided advice for communities opposing municipal and private plants or sites, which came to be known as 'LULUs' or Locally Undesirable Land Use' cases. The campaigners came to be known as 'NIMBYs' or Not in My Backyard' mobilisations. However, the CCHW and scientific experts such as chemistry professor Paul Connet when beyond the local for the scientific evidence which was provided to local communities; information was utilised from Europe, Japan and Australia to provide the data for opponents to toxic plants throughout North America. One significant fact that emerged from this approach was the dramatic increase in dioxin ingestion from the food chain surrounding incinerators. This figure was believed to be as high as 500 times greater than airborne emissions.

The result of the campaigns of the anti-incinerator and anti-toxics campaigns has been increased legislation on emissions and a decline in the construction of incinerators in the United States. The incineration industry has improved its filtering system to reduce emissions in order to address community concerns. However, the anti-incineration movement also contributed a positive initiative to the toxics debate, through its promotion of recycling. The internationally accepted waste hierarchy which locates reuse, reduction and recycling at the top of its pyramid also places incineration and landfilling at the bottom. Paul Connett would also become a lead spokesperson of the 'Zero-Waste' movement, which advocated for the reuse of all materials in production processes.

Campaigns against incineration continue in the United Kingdom, the Republic of Ireland and throughout Europe. In the United Kingdom, a number of campaigns against incineration have occurred in areas such as Guilford and Brent's Cross and Bristol. Anti-incineration campaigns have also emerged in Russia. Campaigners have the benefit of national and international support networks which evolved from the work of the Citizens Clearinghouse for Hazardous Wastes and Professor Connett. Anti-toxics campaigners have also been able to link their concerns with the wider issue of climate change by highlighting the negative impact of toxic emissions on the ozone layer. The internet became a valuable resource for local anti-toxics campaigns that might once have been isolated.

Cultural Accounts of the Anti-Toxics Movement

In 1994, sociologist Andrew Szasz wrote of the methods in which the anti-incineration movement was able to move beyond its NIMBY inceptions in his book *Ecopopulism*. The anti-incineration movement in the Northeast of the United States was the subject of a study by Edward Walsh Rex Warland and D. Clayton Smith titled *Don't Burn it Here*. Both books were instrumental in shaping my own work on the Irish case. The true-life anti-toxics activism of a local mother and legal activist was the subject of a movie titled *Erin Brockovic* starring Julia Roberts in 2000. A movie of the week about the Love Canal campaign titled *Lois Gibbs and the Love Canal* was made in 1982. The issue of communities opposing toxics plants has become a part of contemporary culture, and is regularly featured in television news, documentaries, movies, books and even comics. Like the nuclear industry, the wider toxics and incineration industries have developed a considerable amount of public opposition to their products, and the introduction of innovations or changes are met with a degree of scepticism from concerned communities.

THE IRISH ENVIRONMENTAL JUSTICE AND ANTI-TOXICS MOVEMENT

One unintended outcome of the success of the US environmental justice and anti-toxics movements was the relocation of toxic industries to countries such as the Republic of Ireland during the 1980s and 1990s. One offshoot of this was the attempt to establish a globalised incineration industry through the introduction of 'waste to energy' plants in Europe. Subsequently, the incineration industry lobbied governments in Europe such as Ireland to allow them to provide a solution of their growing waste problems. What followed was the emergence of a number of 'Ecopopulist' community campaigns against incineration across Ireland, in Dublin (Poolbeg), Cork, Meath and Galway. 'Ecopopulism' is the concept developed by the American sociologist Andrew Szasz (1994) to describe the widespread, if not always closely coordinated, phenomenon of community groups coming together to protest against a variety of environmentally damaging threats – has long been a feature of Irish life. For many, changes in Ireland over the last half-century reflect an ongoing disengagement from the traditional patterns of life that had embedded a set of values and practices that allowed rural communities to coexist with their surrounding environment. The grievances of the wider population, which emerged from this perception of a loss of community, have contributed to the growth of populist environmentalism in Ireland, based on 'rural sentiment' – a collective resistance born of the community's inherent understanding of its place in the hinterland, away from the urban centre. During the pre-boom decades of the 1970s and 1980s, such ecopopulism was manifested within the rural-based discourse that characterised disputes over toxic substances and the activities of multinational disputes, among them lead and zinc mining at Tynagh, County Galway, and gold mining near Croagh Patrick, the County Mayo site linked with St. Patrick.

Some feel that strong parochial rural sentiment contributes to the lack of acceptance of official environmental organisations in country-based disputes; as communities attempted to mobilise grievance based on local understandings and relationships, their concerns do not travel well. Such localism has also left many populist environmental groups open to the accusation of being merely cases of NIMBY – the familiar acronym for 'not in my backyard' – when opposition to industrial or infrastructural projects in a community's 'back yard', rather than in a larger systemic context, is identified as the primary rallying point for campaign mobilisation. Populist campaigns often face the problem of translating rural discourses into normal legal hearings. In the Rossport 5/Shell to Sea campaign, for instance, local farmers' attempts to explain that it was unsafe to construct on bogland were simply not understood by those from outside the region, while to the locals building anything on a bog was patently unsustainable. Their point, rooted in rural experience, was reinforced when part of the bog and local cemetery washed away after being disturbed by the development of the refinery site. While specific populist environmental campaigns may wax and wane, the significance of each campaign's contribution to an articulation of community grievance has created a movement of sorts, where outcomes can be measured through an understanding of the extent to which populist fundamentalism has come to be seen as the very basis for traditional rural identities in the post-consumption, post-modern era. The well of grievance that provides much of the underlying discontent for populist campaigns to exploit is the basis for an understanding of exactly how the various environmental campaigns that have occurred over recent decades can be characterised as components of an overall social movement. That movement constitutes part of the social capital of all rural and rural-urban communities. As such, it carries far greater significance than simply the impact of protest campaigns on policy implementation, such as the building of a pipeline, or the construction of a heritage centre in the middle of the Burren. The impact of rural populist discourse goes beyond moments of access to political opportunity structures, and contributes to the shape and nature of the populist Irish political system itself.

Even before the recent economic downturn, communities in Ireland having enjoyed the benefits of extraordinary growth – the so-called 'Celtic Tiger' – had also begun to witness the downside of the economic boom. Infrastructural projects had come to be perceived as threats to locals and their health and environment, or to the heritage of the nation itself. Such offshoots of rapid development as hyper-consumption, a buoyant property market and increased car ownership led to further demand for roads, waste management sites and water treatment facilities. The link between growth and infrastructure was something of a sacred cow: communities that voiced concerns about projects were widely deemed, in the press and in the statements of politicians, to be backward-thinking and against progress.

Recently, we have seen a re-emergence of other ecopopulist campaigns in Ireland, such as the dispute that began in 2005, that of the 'Rossport Five', local farmers who were charged with interfering with a Shell Oil pipeline. Their protests and convictions became something of a nation cause célèbre. Calls for local ownership of local resources were combined with concerns about the risk posed by onshore pipelines as part of the framing strategies of that campaign. One of the most important of the collective responses to development, the 'Rossport Five' emerged in the wake of the attempt to build onshore pipelines on the west coast of County Mayo for the natural gas of the Atlantic Field. Shell Oil had proposed, and been authorised by the government, to build a £600-million high-pressure gas pipeline. Local farmers who lived near the proposed line objected, initially on the grounds of safety, and resisted attempts by the multinational involved to gain access to their land.

As a result, five local men were imprisoned for 94 days for refusing to agree to an injunction that would allow the agents of the multinational access to their property (Leonard, 2006, 2008). What followed was a campaign that gripped the nation's imagination, as 'Shell to Sea' protests sprung up across the island and beyond in response to the community's demand that the gas be processed offshore. The imprisoned men's traditional lifestyle of farming in an Irish-speaking community represented to many observers an authentic and traditional way of life that had been abandoned in the rush to embrace the consumption-fuelled prosperity which had become prevalent in recent years. The fact that the Mayo farmers put family and community above profit or gain provided many observers with an indication of the wider context that had become the subtext of this dispute: the sense that Ireland had, in some ways, sold its soul in return for multinational investment and growth. The men were released in the summer of 2006, and their campaign continues. Despite persistent protests, the construction of the refinery is now nearly complete. The problems of waste disposal related to accelerated growth in Ireland are another flashpoint. One manner in which sustainability campaigns have mobilised is through the use of communications technologies.

COMMUNITY CAMPAIGNS AND COMMUNICATIONS TECHNOLOGIES

One method of gaining increased political access for movements is demonstrated through their increased proficiency with communications technologies. The utilisation of media and internet technologies by environmental protest groups has facilitated a global network which target global corporate and institutional entities. In addition, cyberprotest movements are able to have their mobilisation and framing processes enhanced while the use of internet technologies provides access to the resources of expertise and scientific professional data. These resources can be combined as material evidence in support of a movement's cause, as well as information for media-based framing campaigns. We can chart this growing trend in movement activism by exploring the impact of such the links between Irish protest campaigns and global movements that have provided vital information and strategic material to underpin their campaigns of opposition. One notable feature of the recent political landscape has been the increasing incidents of confrontation between grassroots and elites. These conflicts have occurred in the wake of the declining relevance of the traditional left–right dichotomy, and have been exemplified by the campaigns of opposition led by environmental groups against the globalised corporate sector. We may examine the manner in which new forms of political expression emerge from the environmental movements' utilisation of the new technologies of communication as a strategic tool for campaigns of protest. The use of internet and media technologies by environmental groups has facilitated the growth of a network of committed activists, who provide scientific and technological expertise to like-minded protests around the globe. By exploring movement use of media and internet technologies, this book will outline new approaches taken by grassroots groups as part of their resistance to corporate and institutional actors.

The manner in which communication technologies enhance protest movements by providing leverage and influence for grassroots groups in an era characterised by knowledge flows and technocratic expertise has become a significant aspect of that movement's capabilities. Internet linkages facilitate innovative approaches to political opportunity structures for movements through emergent features of cyberprotest that create a new nexus of capabilities in a globalised age. The 'repertoires and cycles of protest' (Della Porta & Diani, 1999) of new social movements have come to be underpinned by the onset of interactive knowledge flows, networked alliances, improved tactical approaches and advanced mobilisational capabilities through the development of cyberprotest strategies.

The use of internet technology as a basis for support between disparate protest groups is strategically augmented by the utilisation of media networks eager for a steady supply of presentable items on potential ecological crisis for their ceaseless broadcasts. The information highway has become a vehicle for the dissemination of the various components of the environmental movement; academic and scientific expertise, political strategies, legal frameworks and the location of globalised support networks. By facilitating the spread of information which strategically enhances campaigns of protest, the new technologies of communication have become a vital tool in the arsenal of the environmental movements globally. Castells defines internet technologies as a 'privileged tool for acting, informing, recruiting, organising and counter-dominating' (2001, p. 137). Through the application of communications technologies, localised environmental movements can link with similar groups on a global basis, learning from the hard-earned experiences of previous campaigns. This has assisted in the transformation of the once isolated pockets of localised environmental resistance into a worldwide movement able to challenge transnational corporate polluters. An increased reliance on communication technologies by environmental groups has seen a transformation in activists' profiles, changing our perception of the politically dogmatic campaigner into that of a media-friendly advocacy professional with a high level of expertise in a related field. This use of expertise has created what Castells has called 'the new dynamics of social movements' (*ibid.*).

While internet sabotage from computer 'hackers' is an established strategy of the politics of the information age, it is the ability of social movements to mobilise and communicate through information technologies which provides a degree of dynamic innovation to the politics of protest. The internet has become the activist's meeting house in the information era. Networks of opposition have been established and continue to reinforce protests globally, with such success that industries and institutions are now attempting to restructure their own practices in response to the new cultural and political expression of environmental values. This process of redefining cultural values has traditionally been part of old social movement agendas; however, internet technologies make this redefinition or reinterpretation of values much easier for relevant movements today. This shift in social movement capabilities can be seen in the transition of protest groups from a reactive force for change within a dominant ideological paradigm into that of a cultural movement that can sets campaign agendas as witnessed in the ability of social movement organisations such as Greenpeace to influence public perceptions of the environment. In turn, old political conflicts can have their overall boundaries redrawn in this era of communication technologies, as cultural values are redefined through information flows. In this way, old political values (and conflicts) can be resurrected through campaigns framed by the communication of new perspectives.

The restructuring of political and cultural values through the technologies of communication has become an important strategy for social movements in their campaigns of action. It is what Castells refers to as 'mobilisation around meaning' (2001, p. 140). The significance of new political understandings that are shaped by collective action becomes evident in relation to the onset of democratic deficit facing mainstream politics. As parliamentary politics veers towards a centrist, Liberal Democratic monolith, radical political expression had adopted new organisational forms, with the traditional hierarchies of the corporate sector and the state being challenged by grassroots movements which can give equal voice to a multitude of concerned activists, with each able to give as much or as little to a campaign as internet technology allows. Furthermore, social movement campaigns are now structured around the technologies of communication. The medium can affect the message, and often shape it. For instance, a movement's salience may be shaped by the process of media framing at certain key points of a campaign. Public attitudes are swayed by media images, such as Greenpeace running a flotilla in the Irish Sea, alongside ships carrying a nuclear cargo for British Nuclear Fuels. Resolution to the vexed question of nuclear power is not the expected outcome of this form of protest. However, Greenpeace gains a large amount of public support for their movement from the transmission of these images. Ultimately, the media is utilised to create transferable emotiveness around a political event through the creation of a moral frame.

The link between internet mobilisation of protests and satellite news coverage is evident from the events surrounding the Seattle and Genoa anti-globalisation protests. These protests and the massive publicity they received from global media networks has elevated protest to a new level, giving symbolic meanings and outlets for political expression to a new generation of activists. Issues of significance for young people, such as anti-corporatism or environmentalism, can now be forced on to the agendas of powerful groups such as the G8, with an immediacy that bypasses the slow and cumbersome four-year cycle of parliamentary elections. The immediacy of internet communications increases the spontaneous nature of protest events, making them at once attractive to both the casual activist and the media networks which follow these events. The emergence of global networks of disaffected young people has enhanced the anti-globalisation movement's ability to mobilise campaigns of protest. Throughout the last decade a new generation of politically motivated activists with advanced networking capabilities has become part of the discourse of new millennium politics, as characterised by 'McLibel' and Monsanto anti-corporate campaigns (Klein, 2000). Communications technologies are a central component of consumer activism that targets sweatshop production by *Nike* or unsustainable modes of transport like the 'Hummer' sports utility vehicles (SUVs). Cyberprotesters can challenge multinationals that are now recognised as 'the most powerful political forces of our time' (*ibid.*).

For protest movements, the internet allows for a strategic diversity which prevents such movements from becoming stifled by drawn-out campaigns, which can sap morale. Through communication technologies, the battleground of movement protests can be shifted at the press of a keyboard button. This immediacy gives protest movements the ability to link and mobilise with key allies globally. Key events are utilised as mobilisation strategies, with a series of planned tactics ensuring a negative response from the authorities, resulting in the expansion of democratic deficit and moral frames through the filter of any ensuing media coverage. An example of this can be seen in the events around the 2002 'Reclaim the Streets' protest in Dublin. As the police response to the actual march was seen to be over the top, the media coverage and public outcry about police actions on the day lead to a prolongment of the events' salience with the media. The ability to challenge globalised networks of power by protest movements has become an important aspect of the politics of the new century. As traditional institutions are increasingly redundant, new equations of power pit grassroots protest movements against the globalised industrial–military complex. Communications technologies, once the preserve of the latter, have become the weapon of choice for the former. Strategically, the mobilisation, coordination and publicity needed for protest movements at a global level can only be achieved through communication networks. What becomes clear from an analysis of how protest movements utilise communication technologies is the increasing importance of these new forms of technology as a tool for spreading information and organising resistance. The new technologies of communication ultimately create a platform for forms of political resistance that can respond to the needs of concerned citizens in an increasingly globalised world. Essentially, the strategic impact new media forms have had on protest movements has been based on a movement's ability to influence overall media agendas. The ratings potential of protest coverage is dependent on two main themes; the message of the movement, and the potential for an attention grabbing media event emanating from that movement's activities.

Therefore, the subject of movement salience becomes a critical aspect of how media attention is maintained. According to Rogers and Dearing (1988), there are three main features in the issue maximisation process. These are part of an agenda building structure which incorporates factors such as feeding the insatiable appetites of communications elites such as multi-media news corporations through the provision of movement events which embrace the spectacular, and a coordination of media, public and policy agendas. This fusion of demand expectation and need shapes wider understandings of the contentious social events which will maintain public's attention as well as viewer ratings. A central feature of this analysis is the importance of the issue in the context of social expectations. Issue salience is achieved through extensive media coverage, but the issues can only be ignited into activism through the intervention of movement entrepreneurs or gatekeepers. Issue importance is maintained by 'public and policy agenda setting' (Dearing & Rogers, 1996). From this perspective, policy response to an issue is dependent on the importance such an issue is granted by public and media agendas. In this way, successful outcomes to social movement activity are demonstrated to be increasingly reliant on that movement's ability to influence professional media elites, in addition to provoking a public response to their issue of concern. Subsequently, the media's role in a new political equation, as a filter for conflicts between grassroots and elites, becomes a crucial aspect of the new political dynamic.

A further understanding of the centrality of media to the grassroots movements can be achieved through an analysis of what has been termed the 'editorial gatekeeper'. It is the role of agenda shapers to decide what events in society are noteworthy enough to feature on front page or primetime news reports. Their importance to a grassroots movement in need of public support is of paramount importance to the mobilisation process. While political disputes still require the exchange of views from adversarial groups, the influence of media on the public perception of that dispute ultimately influences the outcome of political contestations. What becomes evident from this analysis is the importance of media advocacy and editorial support in order for grassroots movements to gain leverage. In response to this challenge, movements have achieved a greater degree of media professionalism, combined with technological expertise. As the influence of advertisers increases, media coverage and subsequent editorial support can be effectively driven or censored by commercial demand. Another form of external pressure applied to media has been demonstrated by the US government's determined steering of the military action taken in Iraq. In a new departure for war coverage, journalists were embedded with various regiments across the field of operations. Favourable war coverage led to priority postings for the 'embedded' journalists deemed supportive enough of the Bush administration during the Iraq War. In contrast, open criticism of the war led to public accusations of unpatriotic behaviour, and even treachery.

Invariably, the media's coverage of social events can be shaped an underlying ideological perspective. The meanings of societal disputes around topics such as environmental or human rights are subject to an ideologically based conflict of definition. Here again, the media plays a central role, ordaining the public perception of social issues, which is flavoured by inherent ideological demands. One result of this shift in social perception has seen an increased cultural significance of the aesthetics of political disputes, as demonstrated in the campaign of Greenpeace against the French and UK nuclear industries. A further demonstration of the cultural significance of political media coverage can be gleamed from a study of the anti-globalisation phenomenon. This movement has drawn the lines of demarcation clearly, between a grassroots coalition of environmentalists and pacifists, and the formal institutions of global power, such as the G8/G20, World Bank and the International Monetary Fund (IMF). In this way, the anti-globalisation movement allows potential activists the widest definition to associate themselves with certain pro-environment and anti-war sentiments which many people feel some form of empathy. As a result, media coverage is guaranteed, as a mass mobilisation of a broad range of groups from academics to anarchists provides the potential for a 'media event', the focus of which is readily provided for by the heavy-handed response of local security forces.

This represents a problem for both the public governance and the private corporate sectors, as environmental movements move away from social movement campaigns of extreme actions, or violent protest (despite the actions of a some protestors in Seattle and Genoa) and instead present themselves as alternative-minded, environmentally conscious NGOs with a laptop in one hand and a protest placard in the other. Where once violence, at protests or otherwise, from social and environmental movements gave authorities the excuse to respond with the strong arm of the state's military and police apparatus, this aspect of state repression can be avoided by technologically driven protests.

This avoidance of direct confrontation between protestors and the authorities on the streets or at the site of environmental dispute allows for a broader public empathy with the protest movements aims. Furthermore, this strategy of utilising technologies and expertise gives protest movements an air of respectability, which belies their anti-establishment motives, while crucially allowing for increased media access. Indeed, this strategy has become a feature of media processes, as the use of sourced information, conveyed by state-of-the art IT technologies has become a central part of both journalistic and editorial news gathering. This form of political expression through technology utilisation has evolved from the unease felt by local communities in the face of the loss of powers at local and national governmental levels, and the rise in multinational globalisation. As the power of the nation state recedes, communities are utilising the new technologies of the information age, and combining them with specialised areas of expertise, to directly oppose corporate entities or toxic industries.

'Ecopopulist' campaigns therefore represent a new form of protest, one which differs from previous environmental disputes in Ireland. Essentially, ecopopulism can be seen as part of what Castells had described as 'an array of transnational social forces animated by environmental concerns rather than a protest group rooted in either NIMBY perspectives or deep green ideology. As such, ecopopulist campaigns can be linked to a worldwide anti-corporate movement which having environmental concern as a basis for their thinking, has as much to do with addressing the onset of risk society and democratic deficit as any emergent eco-consciousness. In the Irish case, this crisis is represented by the onset of a consumer society and a reliance on multinational investment that has led to an industrial pollution and a waste crisis for society. Despite an increased economic output, many communities in Ireland have begun to experience a sense of alienation from this 'Celtic Tiger' economy, as multinationals have imposed on both the environment and quality of life of many citizens.

Moreover, the political processes, both in Ireland and internationally, have been altered to meet the demands of this new form of politics. As the campaigns of ecopopulists have demonstrated, this new form of politics is based on both the fears of local communities in the face of the growing power of the multinational waste management sector (of which the incineration industry is but one component) and a growing democratic deficit presented by the weakness of the state in responding to its citizen's fears and needs in regard to both waste management, and the subsequent health risks they may pose. Ultimately, the citizenry has responded to this challenge through protest movements, circumventing established political channels, and turning the new technologies of the corporate sector on the corporations, to oppose environmental degradation. When viewed from an international perspective, this new movement of political resistance has become both an important arena for political expression, and a new form of political action.

This book examines some of these issues associated with disputes around the bourgeoning waste crisis, a consequence of increased consumption associated with accelerated growth, led to community action. Throughout the boom years of the 1990s, Ireland had fallen foul of European directives on waste management due to an over reliance on landfill, which received up to 93% of the country's solid waste. The Irish Environmental Protection Agency was set up in 1992 to deal with such issues, but many critics charged that it quickly became associated with business interests and government appointments, rather than with environmental protection. This book primarily focuses on the case of the growing west of Ireland city of Galway, and follows the political cycles which occurred alongside the social movement mobilisation of a community campaign for sustainable living.

Dr. Liam Leonard
Galway, June 2011

CHAPTER 1

INTRODUCTION

1.1. MOBILISING RESOURCES, EXPLOITING OPPORTUNITIES AND FRAMING ISSUES

Community campaigns for sustainable living emerged in the Republic of Ireland during the 1990s, in response to the state's introduction of regional waste management plans that incorporated infrastructure such as 'waste to energy' incinerators and 'superdumps'. This book examines the manner in which such groups mobilised internally as well as exploiting the resources and political opportunities that emerged externally during the course of their regional campaigns. In particular, the book focuses on how these community campaigns mobilised resources and framed their challenges through the exploitation of the political opportunities provided by the state's policy agenda, and later through the state's response to the protests that followed. Other issues examined in this study include attempts to influence the manifestos of the political parties contesting the 2002 general election, as part of the political opportunity cycle.

This sociological study is set at a time, between 1999 and 2004, when community groups in the Republic of Ireland were opposing the waste management policies of the state from an initial NIMBYist or 'not in my backyard' perspective. However, the book demonstrates the extent to which such campaigns moved beyond this narrow inception and became part of a wider political agenda that linked communities nationally and, even to a lesser extent, internationally. The group at the core of this study, Galway for a Safe Environment (GSE) emerged at a time when the state was exposed to 'bottom-up' NIMBYist challenges due to growing consumption and waste in the post 'Celtic Tiger' years of economic growth, combined with the 'top-down' pressure from EU waste directives and increased mistrust of politics in general in an era of political scandals and tribunals of enquiry.

This investigation sets out to provide understandings of how this regional community group mobilised internal resources such as expertise and exploited external political opportunities. It examines the framing of the anti-incineration issue and discusses the reasons why community groups would attempt to undertake a wider campaign of political ecopopulism and the unforeseen problems that resulted from this attempted change in direction.

1.1.1. Theorising the Study

The study uses three main theories to illustrate these points. The theory of 'the four most salient changes to the dimensions of political opportunity' provided by Sidney Tarrow in his *Power in Movement: Social Movements, Collective Action and Politics* (1994, pp. 85–90, 1998, pp. 75–80) utilised to provide an understanding of how a model of political opportunity can be built from the key moments of a campaign. This concept provides an understanding of what gives a movement access to political opportunity structures and what impact alliances play in the course of a campaign.

These four salient changes of 'increased access', 'divided elites', 'unstable alignments' and 'influential allies' (*ibid.*) will be applied to key moments of the campaigns, to provide further understanding of how a campaign can gain temporary moments of access to political opportunity structures.

The manner in which 'increased access' was gained by exploiting the external resources of political opportunities, such as the state's omission of the potential health risks in their waste plan, the councillors' vote on these regional plans, and most significantly, the 2002 general election will be outlined. The community advocates also exploited the existence of the 'unstable alignments' and 'divided elites' that existed between the state, the local councils and the consultants for the plan, will be outlined. Campaigners were able to forge links with 'influential allies' such as Dr. Paul Connett, the noted American anti-incinerator campaigner, as well as local politicians from the government coalition and from opposition parties throughout their campaign.

The study also utilises Snow and Benford's theory of 'Master Frames' (Snow & Benford, 1992, pp. 133–150) to provide understandings of how movements frame issues during the 'cycles of protest' (*ibid.*). Snow and Benford provide a framework with which the process of how the framing of meanings can be understood, through a process of 'amplification', 'transformation' and 'generating' meanings throughout a campaign (*ibid.*).

An analyses of the three 'main frames' of health risks, NIMBYism and democratic deficit is presented, to chart the progression of regional community campaigns. These regional movements emerged from single issue anti-incineration campaigns into a national 'ecopopulist' challenge to the perception of democratic deficit in the state's waste policy and in its response to the campaigns is part of the study. An understanding of the relationship between the mobilisation of internal resources, the exploitation of opportunities and the framing of key issues during these campaigns is provided. The study also provides evidence of where this process gave 'increased access' to political structures, or where conversely the 'perceived' opportunity restricted access to the political structures surrounding this issue. In so doing, the thesis demonstrates both the opportunities and the constraints of community campaigns for sustainability, while simultaneously demonstrating the strengths and limitations of Tarrow's 'four most salient changes', such as the limitation of community social movement's alliances with elector candidates in national politics. This 'perceived' opportunity for 'increased access' through an 'influential ally' would have unforeseen impact for community campaigners, as promises made locally weren't delivered on nationally.

Further understandings of this process will be achieved by exploring the mobilisation of key campaign resources such as expertise and networks by the campaigns, demonstrating how the process of mobilising internal resources provides 'increased access' to political opportunities at key points of the campaign. During these key moments, a 'perceived' opportunity may increase access or create problems for a campaign or movement. This study highlights the key moments of real or perceived opportunity for movements during their campaigns, indicating whether these opportunities increased access or created further constraints for such movements.

Szasz has detailed the US anti-toxics movement in his text *Ecopopulism* (1994) and argues that campaigns can move away from their NIMBYist inception into ecopopulist politics. The study also provides an explanation of why regional campaigns were unable to make the transition 'from NIMBYism into radical ecopopulism' (Szasz, 1994, p. 69) during its campaign. It also asks why they were unable to make this transition, despite their attempts to politicise their campaign, although it highlights the attempts these movements did make to move from their NIMBYist origins through its attempt to create wider networks and alliances in an attempt to become more 'ecopopulist' in the understanding of this term set out by Prof. Szasz in his work.

1.1.2. Internal Resources and External Resources

Tarrow (1994, p. 84) provides an understanding of the significance of how movements mobilise internal resources such as the expertise of key personnel and leaders, and how such movements exploit the external resource of political opportunities. The study highlights the manner in which a grassroots environmental movement in Galway (*GSE*) mobilised the internal resources of political and scientific expertise in order to undertake a challenge to the government's waste policy that included regional incineration plants. GSE's campaign is shown to have occurred during the second of two main phases of environmental campaigning in Ireland. These two phases occurred in the years before and after the economic boom of the 'Celtic Tiger' years of accelerated growth in Ireland. The first phase, from 1970 to 1990, occurred during the state's prioritisation of multinational-led industrial development. Evidence is provided to demonstrate that environmental campaigns during this time were locally focused and opposed multinationals that were causing pollution.

The book also demonstrates how the second phase of environmental campaigns occurred in the aftermath of the establishment of the Environmental Protection Agency (EPA) and a waste management framework. The study also asks why it was that environmental campaigns during this second phase focused on opposing waste treatment plants and incinerators, as well as landfills, in the aftermath of the post-boom increase in consumption and waste creation. GSE's campaign is shown to have occurred in this second phase and illustrates the manner in which certain characteristics, which were innovatory for an Irish environmental movement, such as its utilisation of communication technologies to further its aims, as well as its ability to create networks with US and Irish anti-incinerator activists. Irish environmental literature is examined to highlight the significance of GSE's case within the context of existing studies of Irish campaigns by writers such as Cox (2001) and Garavan (2004). This context provides an understanding of the significance of GSE's utilisation of expertise networks and use of internet technologies as part of their attempt to link with anti-incineration networks in Ireland and in the United States. The study illustrates the manner in which GSE politicised their campaign during the course of a general election and the mixed results this had for the group. While most Irish environmental movements have based their strategies in legal challenges and oral hearings (Taylor & Flynn, 2003, p. 119), the study shows that GSE were able to use the global networks of their US allies to challenge the state's scientific data on incineration, making the potential health risks, and undemocratic approach of the government on incineration, political issues during the 2002 general election.

My investigation utilises Snow and Benford's (1992b) discussion of 'Master Frames and Cycles of Protest' to build an understanding of the frameworks that surrounded GSE's exploitation of the political opportunities that emerged during the campaign. In particular, I show in Chapter 5 how the three main cycles of GSE's campaign were shaped by GSE's framing of three crucial issues. In the first cycle, GSE highlighted the health-risks posed by the dioxins and furans contained in the emissions from incinerators.

Further issues examined include the connection between the first two frames of health risks and campaign NIMBYism and outline their impact on the third frame of democratic deficit in the electoral cycle. The study details the process in which GSE 'amplified' and 'transformed' meanings around the three frames, to create 'new meanings' (Snow & Benford, 1992b, p. 136) around the incineration issue and wider political events. It also highlights the significance of 'interest driven' (Grove-White, 1993, p. 22) perspective on the potential health risks of incineration, as GSE embarked on a campaign of public protests, meetings and a petition, as well as undertaking a media campaign. The significance of the media campaign that involved local and national media, with GSE utilising the Internet technology to provide 'interest'-led data sourced from the US anti-incineration expert, Dr. Paul Connett will be pointed out. This media campaign is shown to have allowed GSE to involve national television and print media in a local issue.

This study asks the question of why GSE did not make the transition 'from NIMBYism to ecopopulism' (Szasz, 1994). From the inception of their campaign, GSE were concerned about being portrayed by their opponents as merely self-interested NIMBYists, a fate that had befallen previous environmental campaign in Galway. The study shows how GSE created networks with Connett in the United States, as well as other anti-incinerator campaigns in Meath and Cork. This process involved the expansion of the NIMBY frame from a community group in an area on the edge of Galway, into a type of national network incorporating the Galway, Cork and Meath campaigns. This national network is shown to be the beginning of a nascent anti-incinerator movement, with GSE providing interest-led data from Connett to the other NIMBY groups in Ireland.

Another issue examined is GSE's exploitation of concerns about democratic deficit surrounding the incineration issue. To construct this frame, GSE made use of the political opportunities that came its way throughout the campaign. These political opportunities included the omission of potential health risks from the CWP draft, the use of 'interest'-led data to influence the city councillors' vote against the CWP, and the opportunity of the 2002 General Election. During the election campaign, GSE's third frame enveloped the primary concerns about the health risks of incineration, as GSE's tactic of influencing the parties became a campaign against the Fianna Fail party as an extension of the democratic deficit frame. This tactic would fail in the aftermath of Ireland's then largest party Fianna Fail's electoral success, leaving the campaign facing strategic difficulties.

The overall political opportunity structure of the case will be shown to have evolved on three main levels. The study demonstrates these structures that included the local level, which included GSE and the relevant local authorities in Galway; the national level, which incorporated the coalition Fianna Fail and Progressive Democrat (PD) parties, as well as their corporatist partners from the business lobby and trade unions; and the wider transnational level, where the environmental institutions of the European Union (EU), whose directives on waste management created certain pressures for local and national government in Ireland, are relevant.

1.2. BACKGROUND TO THE CASE

Ireland's phase of accelerated economic expansion in the late 1990s has been characterised by patterns of increased consumption and waste production (Fagan, 2003). One result of this has been the emergence of a waste management crisis, as levels of waste outstripped the capacity of the existing waste management system, which had mainly relied on landfill. State policy began to focus on alternatives to landfill such as incineration, in response to EU directives, while grassroots community groups mobilised in opposition to the state's waste management policy. GSE came into being after the announcement of plans for the use of incineration in the draft *Connacht Waste Plan* in September 1999, as part of the government's regional waste management policy. GSE's committee began to implement a campaign of opposition to the *Connacht Waste Plan* through the strategic highlighting of the potential health risks posed by incineration emissions.

The main features of GSE's campaign are highlighted, including the importance of its committee that emerged to provide leadership. GSE's links with the US anti-toxics movement were empowering to the extent that it was able to acquire the persuasive scientific arguments that underpinned their campaign. This internal resource of scientific evidence provided the group with a solid platform with which to mount a challenge to government policy, particularly as the regional waste plans neglected to mention any of the potential health hazards posed by incineration.

The study also looks at the question of the relationship between the state and the industry, and how these relationships were viewed by community groups like GSE. The other main participant in the dispute, the state, was seen to be less than forthright in its approach to the waste management issue. This allowed GSE to exploit the public's concern with a growing democratic deficit in Irish politics, particularly in the wake of the many scandals and tribunals of inquiry that had come to bedevil the political scene. Essentially, GSE's campaign can be identified as having two main external resources. The primary external resource was the inclusion of incineration as part of the state's waste management agenda. A second external resource can be identified around GSE's attempt to gain access to the state by engaging with the electoral system during the 2002 election in an attempt to recruit influential allies. This engagement included a campaign against Fianna Fail candidates in that election and support for Green Party and the now defunct PD candidates.

The question of how resources such as technologies of communication were mobilised as part of GSE's strategic approach is another area of interest in this campaign. One of the ways GSE mobilised internal resources was through its effective use of media and internet technologies to spread their point of view on incineration. This utilisation of media reinforced their links with the US anti-toxics movement, while also facilitating productive interaction with television, radio and press sources, both locally and nationally. By providing a strong media profile for their campaign, GSE was able to mobilise support for their various strategies, which included the gathering of a petition, the holding of public meetings and debates on the issue, and the staging of public protests and marches. I examine GSE's use of tactics and mobilisation of support for its campaign, by placing these approaches in a resource mobilisation (RM) framework, as well as outlining the manner in which the mobilisation of internal resources allowed for greater access to the external resources of the political opportunity structure.

My methodology utilises RM theory to examine GSE's campaign due to the established nature of RM theory in the study of movement campaigns. The RM theoretical approach looks at the utilisation of internal resources such as leadership and networks, prior campaign experience and expertise. These resources allowed GSE's committee to establish itself as movement 'entrepreneurs' (Della Porta & Diani, 1999, p. 9). By contesting the waste management issue, GSE's committee was able to gain access to the political opportunity structure (POS) in ways other Irish environmental movements had not previously achieved and in a manner that involved 'the politicisation of everyday life' (Taylor & Whitter cited in Morris and McClurgh Mueller, 1992, p. 12). This study demonstrates the manner in which the 'politicisation' occurs in GSE's case, as environmental issues have come to be a feature of modern life.

The key focus of GSE's use of scientific data highlighting the health risks posed by incineration was aimed at challenging the approach of the state to the waste issue, which had ignored the health risks associated with the mass burn incineration in their regional waste plans. This study highlights the centrality of the expertise that underpinned this challenge, an expertise that allowed GSE to challenge policy through the domestic political process. This use of scientific expertise by GSE influenced the rejection of incineration by the Galway city council. 'Interest-led' scientific data are shown to be an important part of GSE's challenge to the state's data on health risks, in an era when scientific data from challengers and authorities is contested in public debates.

I also incorporate a discussion of how the creation of links at local, national and international levels allowed GSE to operate within a wider anti-incineration network. The prior campaign expertise of GSE's leadership was a crucial factor in GSE's ability to create new meanings from their three frames. GSE's committee overcame this potential difficulty by forming significant alliances and by providing detailed analysis based on their interest-led science.

My research on this case goes on to examine the importance of such networks and alliances. GSE was shown to have forged alliances with other campaigns against incineration, particularly those in Cork and Meath. In so doing, GSE was able to provide interest driven scientific and political expertise to these other groups. This provision of campaign support allowed GSE to become a resource for the wider community, something that distinguished it from the type of NIMBYism displayed by first phase anti-toxics groups in Ireland. By providing external sources of expertise with internally mobilised arguments fashioned for the Irish case, GSE was shown to have raised the profile of the Irish anti-toxics movement, and overcame traditional difficulties faced by previous Irish environmental movements, which it has been argued had grown sceptical of UK- or EU-based expertise (Tovey, 1992, p. 6). On an international level, GSE's links with the US anti-toxics movement activist Dr. Paul Connett was crucial to its challenge to state policy. The use of communications technologies provided GSE with a window of opportunity to put forward its views, at crucial times in its campaign against incineration. It has been argued that certain environmental movements can become the 'moral entrepreneurs' of environmental issues, 'promoting public concern and organising crusades over issues which then gain public prominence' (Tovey & Share, 2000, pp. 513–14). This study demonstrates the significance of moral entrepreneurs such as GSE to the contestation of environmental issues that is becoming an increasing part of this 'politicisation' of the public sphere.

This study examines the social cleavages that occur as a result of this contestation of science. The fact that GSE could use 'interest driven science' to challenge the state's 'official science' (Grove-White, 1993, p. 22) provided it with an opportunity to shape the 'orthodox moral discourse' (*ibid.*) surrounding the issue of incineration. Here GSE's campaign had some impact, as it was able to mobilise the internal resources of networks, expertise and communication technology, to create a degree of understanding about the potential health-risks posed by incineration, as well as the negative impacts that health-risks would create for Ireland's agricultural and tourism industries. The opportunities that emerged from this mobilisation of resources provided GSE with the access to political structures necessary to expand their campaign of opposition to the state's waste policy.

The book also investigates internal and external resources, as part of Tarrow's (1994, p. 85) understanding of the interplay between mobilisation of resources and exploitation of political opportunities. Having introduced the internal resources mobilised by GSE, the study goes on to consider the external resources, based on political opportunities, surrounding the response of the state to GSE's campaign. Despite the requirement for public consultation contained within the government's regional waste plans, it became clear that the state's response to this form of challenge would be less than tolerant of debate.

These external resources are shown to be derived from the state's approach to the waste issue. GSE anticipated the pivotal importance of Galway City's acceptance of the waste plan for the West of Ireland, being the main urban centre of the region. This is shown to be the case, as the proposed incinerator for Galway was to receive waste from across the West of Ireland, a fact that made the county council's rejection of the *Connacht Waste Plan* so crucial. The study shows that as GSE's campaign gained greater momentum Galway's councils rejected the planned incinerator, throwing the state's regional waste policy into disarray.

The case demonstrates the manner in which political opportunities can be exploited by an environmental campaign that mobilises internal and external resources. Through its mobilisation of expertise, technology and networks, GSE was able to challenge the scientific data contained in the state's waste plans. This allowed the group to contest the official position on the issue, and led to a 'confrontative' (Sharpf, in Kriesi, 1995, p. 174) response by the state, according to GSE. This book illustrates how this undemocratic response led to the emergence of a further political opportunity, which allowed GSE to use the incineration issue in the 2002 general election. Ultimately, the success of the government parties in that campaign led to GSE's decline, although the state has not yet introduced incineration in Galway, despite granting incineration licences in Cork and Meath.

The study also provides an account of how state policy was shaped by its commitment to industrial growth and by the requirements of EU directives, which framed its waste management policy within an 'ecological modernisation' (EM) (Christoff, 1996) framework. The requirements of EM are a central feature of the 'Europeanised' (O'Hearn, 1999; Taylor & Flynn, 2003) approach to waste management in the Irish case.

1.3. CASE STUDY METHODOLOGY IN SOCIOLOGICAL RESEARCH

There are various types of case studies, including 'exploratory', 'explanatory' and descriptive' (Yin, 1994); or 'intrinsic', 'instrumental' or 'collective' (Stake, 1995). Exploratory cases examine the background to cases, often as a prelude to further studies of a given topic. Explanatory cases take a causal observational approach to the gathering of documentary evidence, while descriptive cases provide a descriptive theory that is developed to contextualise the study. Case studies are in-depth investigations, as opposed to sampling research, according to Yin and Stake. However, particular cases are chosen to allow for the maximum analysis within a particular time frame.

The research must identify the unit of analysis, creating a dialogue between the research and the participants in the case being analysed. In this way, the case study can provide a counterpoint to 'elite perspectives' that are often prevalent in sociological studies (Feagin, Orum, & Sjoberg, 1991). Case studies can be used to present triangulated research, combining theoretical concepts with investigative data (*ibid.*).

Triangulated research strategies provide explanations for social events, using multiple data sources, providing validity and meaning for the research that had been undertaken. In my research, I have combined data, theory and methodological triangulation with the investigative approach of longitudinal research. My collection and analysis of data took place over a three and a half year period, between January 2000 and August 2004. Changes in the degree of access in the political opportunity structure for GSE were measured throughout this period, as was the mobilisation of resources by GSE. Patterns of change in the fortunes of GSE were described and explained against the backdrop of the responses of both the EU and the state to the waste crisis in Ireland.

My research proceeded in three stages. The first of these was the collecting of data through participant observation in the GSE campaign. This stage of the research included attending public meetings, marches, meetings of the Galway City Council, and public debates that were relevant to the case. The second stage included interviewing participants in the campaign, as well as the gathering of secondary sources such as press reports of GSE's campaign. These secondary sources were, for the most part, drawn from the local print media in Galway. In addition, I was able to attend almost all of the events concerning GSE's campaign over the three and a half years of the study. This first-hand knowledge meant that I was better informed of the events than the Dublin-based national media. I had also been involved in local politics and was able to obtain information directly from participants on both sides of the debate. This, together with the information provided for me through interviews or casual conversations, allowed me to build a comprehensive understanding of GSE's campaign. However, my involvement in local politics may have led to some researcher bias, although I consciously attempted to avoid any outright bias in the course of research. Yin (1994) has presented a fourfold model for the application of a case study. These include the explanation of the linkages in social events; the description of their contexts; an account of the event itself; and the evaluation of the outcomes of this event. This research has applied these four approaches to the case study of GSE's campaign. The relevant links were established between GSE and like-minded anti-incineration groups in the United States and Ireland. A review of comparative cases allowed the GSE case to be placed in a context of anti-toxics campaigns in Ireland. The outcome and implications of the case are also analysed.

The strengths of participant observation are demonstrated in my study. This approach provided accounts of the key events of GSE's campaign. Through participant observation, I was able to build up a good understanding of aspects of the case that may not have otherwise become evident. These aspects include the extent of GSE's network of alliances, their tactical choices to avoid appointing outright spokespersons, and their attempts to influence the 2002 general election, all of which represented a new approach for an Irish environmental movement. Validity and reliability has been established by using multiple sources as evidence, within a case-study approach as suggested by Yin (1994). External validity is provided by an interpretation of these findings within a theoretical framework. This examines environmental social movements in the context of RM and political opportunity structure theories, while analysing the state within the context of its relationship with the institutions of the EU. The manner in which the state and movements are influenced by Ireland's political structures, which are shaped by neo-corporatism and clientelist relationships, is also discussed, as these political structures impacted upon the case as GSE attempted to gain access during the 2002 general election.

1.4. STRENGTHS AND WEAKNESSES OF A CASE STUDY APPROACH

In conducting the case study, it became apparent that there were a number of strengths and weaknesses that manifested themselves during the period of research. Yin (1994) has identified six sources of evidence available to the case study researcher. These include 'documentation'; 'the use of archival records'; 'interviews'; 'direct observation'; 'participant observation' and 'physical artefacts.' I was able to use the first five of the six, combining the data and findings of each source of information to address the case. Yin (1994, p. 80) has established the strengths and weaknesses of using these types of evidence. For instance, my documentation provides a 'stable and repeated review of data' with an exact coverage over a particular time frame. However, this evidence may contain certain biases within the reporting of events, either of my own or of the press coverage I have surveyed.

While my use of media was largely focused on the local print media in Galway, I was able to cover primary and secondary sources from across Ireland, as well as being able to utilise GSE's pamphlets, documents, press releases and their website. The websites of Irish and US environmental, anti-incinerator and pro-recycling group were also consulted. Nonetheless, these sources could also be said to share certain biases, as was true of the official press releases, websites and policy documents of the state, which I also used for this research. I selected relevant personnel whom I thought to be the most centrally involved for interview. However, these interviews may present certain biases due to the questions asked, or the responses given, in light of the contested nature of the incineration dispute. Direct and participant observation allowed me to cover the events as they happened, and place these events in an overall context. However, these observations took time, and were dependent on my own ability to gauge their importance at the time. On balance, the observation I conducted allowed for a greater insight into events than an over-reliance on documentary evidence might have provided. The combination of both observational and documentary evidence allowed for 'a chain of evidence' (Yin, 1994) to be maintained.

The fact that I had some local political experience, and knew some of the participants, both in GSE and in local politics, provided me with a degree of access to the main protagonists in the case, as well as allowing me a certain insight into the wider political events that were occurring around this dispute. This involved fieldwork, participant observation and interviews with the relevant activists involved with GSE's campaign, as well as local and national political figures.

My research made extensive use of local press reports of the campaign, drawing from reports from the local radio station '*Galway Bay FM*' in addition to reports from *The Connacht Tribune, The Galway Advertiser, The Connacht Sentinel* and *The Galway Independent*. These local sources were favoured over the national press that only provided sporadic summaries of the main events of the case. Local sources provided much more detailed coverage of GSE's campaign. On a national level, GSE's ability to utilise expertise and technology led to that groups influencing the broadcast media's reporting of the case. The main evening news of the national broadcaster, Radió Telifís Éireann (RTÉ) featured live interviews, reports and debates with key political actors and prominent members of GSE's committee. These reports included a live debate between GSE spokespersons and national political figures, which led many observers to believe that GSE was a movement of substance.

Participant observation was central to this study. The principal committee members of GSE were contacted about the nature of this research, and I was placed on the group's mailing list. This allowed me to attend all GSE's protests and meetings. The group's website was also utilised for research purposes and proved a useful vehicle for finding information and press releases from the group. Another point that facilitated the research was the academic background of some of the principal actors in the campaign. The key members of the group were quite open in providing me with information, something that facilitated the study.

Throughout the course of the study, I used selective sampling, interviewing the key people in the dispute. My ability to interact with the key participants provided me with the opportunity to interview Professor

Paul Connett, who has been described as an 'identifiable' single individual symbolising the opposition to incineration industry (Walsh Warland & Smith, 1997, p. 17). My interviews also brought me into contact with local politicians and national political figures such as Michael D. Higgins, the president of the Labour Party, who had met with the leaders of GSE at an early stage and had provided that group with strategic advice on creating events for the media to cover. The participants were chosen due to their prior experience with environmental movements, their ongoing involvement in environmental campaigning, and due to the fact that they were part of an overall anti-incineration network. Once I had made contact with GSE's committee, I was assisted in my attempts to contact other participants.

I was also able to attend closed meeting of Galway City Council, having been given permission by the then City Manager, Mr. Joe Gavin, after I had interviewed him for this study. This allowed me a privileged perspective on the debates taking place in City Hall. As such, a wider political understanding of the case could be established, which is congruent with an issue history approach that attempts to construct 'the full history' (Szasz, 1994, p. 163). This study undertakes to present a case history of GSE's campaign, while placing it within a theoretical framework that can reveal its crucial features, and areas of distinction, in a broad analysis. It examines the achievements and problems that the politicisation of an environmental dispute creates for an environmental movement, as the shifting dynamic of the political opportunity structure surrounding the case provides groups like GSE with moments of access to the political structures of the state, without providing these groups with full access to the policy agendas which they wish to influence.

CHAPTER 2

THEORETICAL FRAMEWORK: MOBILISING INTERNAL RESOURCES AND EXPLOITING EXTERNAL OPPORTUNITIES

2.1. INTRODUCTION

This chapter will establish the two main strands of the study's theoretical framework. These strands represent the internal and external resources in the GSE case. The resource mobilisation (RM) strand refers to the internal resources of GSE, while the political opportunity structure (POS) refers to external resources. By referring to the literature on social movements particularly that which dwells on RM and the exploitation of political opportunities, the study will provide an understanding of collective action. This study investigates the campaign of an environmental movement that challenged the waste policy of the state, on the issue of incineration. As the state changed waste policy, the Galway campaign mobilised internal resources and exploited external political opportunities. The shifting nature of this opportunity structure may affect the patterns of internal mobilisation, utilisation of resources and types of networks a movement implements. Government responses to such challenges may also influence the patterns of collective action, as movements attempt to exploit the opportunities of the wider political environment.

The first section of this chapter examines social movement theory, distinguishing new social movements (NSMs) from their classical or old counterparts. Types of environmental movements are then outlined, to facilitate an understanding of the ideological features of the case. The significance of middle-class participation in environmental movements will also be addressed. The manner in which environmental movements mobilise resources will then be detailed, while the activation of networks and alliances will be paid specific attention. The section will also discuss how the features of RM that emerge in a given case may facilitate attempts to gain access to the POS. Political opportunities will be discussed in relation to the degree of access or non-access that exists around the political institutions at local or national level, thus creating shifts in the degree to which types of opportunities can be utilised during periods of collective behaviour.

Among others, Martell (1994, pp. 111–118) has identified a dichotomy between 'old' and 'new' social movements. Old movements follow formal patterns of organisation, with hierarchical structures and elected leaders. Change is attempted by engaging with the formal political institutions. NSMs have emerged from the rights campaigns of the 1960s. Student politics, peace protests, feminist discourses and environmentalism became prominent, while human and civil rights groups came to the fore among the NSMs. Many social movements are based on grass-roots networks, favouring protest and direct action as their methods of strategic action. However, NSMs have also tended to adopt professionalised structures, providing expertise and skills in areas such as science, politics and communications.

According to the pluralist approach, social movements use collective behaviour as a means of 'bargaining to achieve goals' (Gamson, 1975, p. 133). The goals that movements strive to attain may stem from shared grievances held by a social group. These grievances can be used to create consensus among adherents, who then mount a campaign around issues surrounding this grievance. The motivation for NSM activity stems from grievances held by individuals or communities regarding increased intrusion by state and market forces into areas of personal or community life.

According to Melucci (1996), rights issues, surrounding the reclamation of autonomy from the state by new middle class professionals has supplanted economic class struggle as the main motivation for social movement participation. As such, NSMs are concerned with effecting a restructuring of state and society, reconciling collective interests, increasing autonomy from the state's agenda of developing infrastructure and changing the nature of state intervention. NSMs are defined through their activist base, whereas formal political parties have a traditional ideological basis that defines their politics and participants from the perspective of left or right wing political allegiances. While social movements are concerned with a shared reconceptualisation of social relations, mainstream political parties maintain a hierarchical and centralised approach to political and social arrangements; apart from the Greens, who have a more informal structure. However, social movements are dependent on the wider political environment for opportunities, while governments may tolerate or even encourage movements to channel public dissent.

2.2. ENVIRONMENTAL SOCIAL MOVEMENTS

Environmental social movements have emerged since the 1960s as concerns about the effects of ecological degradation have increased. These movements have challenged concepts of industrialised growth that have dominated political thinking over recent centuries. Although environmental groups can network with one another, exchanging expertise and support, the localised focus on environmental grievances can lead to accusations of 'NIMBYISM', or the 'Not In My Backyard' syndrome. NIMBYism entails local populist opposition to sitings of toxic industries or waste plants in localities or neighbourhoods. NIMBY groups frame the 'moral discourse' (Grove-White, 1993) surrounding the environmental and health risks facing communities where toxic plants are situated, and highlight the potential economic and health costs which may result from the distribution of toxic effluents and emissions. These NIMBY groups are characterised by intensive outbreaks of local activism, as public responses are galvanised in opposition to hazardous plants. Protests are used to bargain for the restoration or maintenance of collective goods such as clean air or waterways.

Environmental organisations may be dependant on the goodwill of external agencies for other resources, such as financial contributions or favourable media coverage. In order to attract such support, environmental organisations depict themselves in a manner that can exploit the wider sympathies of a public that may be supportive of environmental issues without wishing to become participants in a campaign. In this way, environmental groups may exploit wider public concern for the global commons, given that shared environmental goods such as clean air or food products invoke a degree of concern across society. Accusations of NIMBYism may therefore be overcome, as environmental movements present themselves as responsible protectors of the environment. Environmental movements may undergo an 'ideological development' (Szasz, 1994, p. 77), as increased professionalism, wider networks and political interaction create understandings of how environmental issues overlap at a national or global level.

Environmental protests are organised by networks that exploit resources and opportunities. The distinction between 'old' and 'new' social movements can be located within this understanding of a social movement sector that provides resources and negotiates opportunities and mobilises campaigns. The form of organisational structure employed by environmental movements may vary, ranging from the 'participatory, antihierarchical and anti-institutional ...'on the one hand to 'formalised, hierarchical and oligarchic organisations' on the other (Rucht from Klandermans, 1989, p. 63).

The variance in environmental movement organisational structures is replicated in the different types of environmental movements that organise challenges against political and scientific orthodoxy. Rucht suggests three distinctive types of environmental thinking that shape movement ideology. These are built around the following:

1. Conservationism; the aesthetic, ethical and religious protection of nature. Conservationist methods include respect for the rights of nature, preservation of natural space or parks and campaigns of aesthetic education.
2. Environmentalism involves a combination of concerns for quality of human life and of the natural environment. Environmentalism embraces scientific, economic and political arguments about the policies that impact upon lifestyles and the ecosphere.
3. Ecologism is concerned with a holistic or utopian conception of human existence in harmony with nature. Ecologism prioritises the concerns of nature over human needs, advocating changes to existing lifestyles and political systems which should reflect the primacy of nature (Rucht from Klandermans, 1989, pp. 64, 65).

The literature points to a dichotomy between ecology-centred (ecocentric) and human-centred (anthropocentric) types of environmental movement. One aspect of ecocentric thought is concerned with a critique of technology. O'Riordan (1989, p. 9) has identified another dichotomy, between what he refers to as 'ecocentricism versus techno-centrism'. Techno-centrism focuses on reforming technology to prevent some aspects of environmental degradation, while an ecocentric focus in contrast advocates the complete reform of socio-political patterns by giving priority to the environment. In this way, politics may be viewed from an environmental perspective. O'Riordan (1989, p. 9) outlines the techno-centric 'belief in the retention of the status quo in the existing structure of a political power' associated with middle ranking executives and environmental scientists. These people have 'faith in the application of science and market forces' (*ibid.*) and feel that institutions can adapt to environmental needs. Techno-centric approaches are at the centre of an 'ecological modernisation' approach, which sees environmental pragmatism as an efficient part of economic and industrial processes.

Martell (1994) examines the general conditions under which environmental movements emerge. Movements may have a shared or 'collective' interest, and pursue goals which will challenge or change institutions, without operating through the channels of formal party politics. As the institutions of the state fail to deal with the environmental grievances of communities, movements are organised around issues of local concerns. Indeed, due to the state's primary focus on infrastructural development and competitiveness, state agencies are often the target of environmental movement campaigns. If the political process is seen to be exclusive, and policy that has an ecologically harmful aspect is seen as imposed undemocratically, then movements can form to express alternative or oppositional positions. Scott (1990, p. 145) has provided examples of neo-corporatist government arrangements that exclude environmental concerns from the political agenda. As political parties, industries and trade unions map out their policies, environmental issues may be overlooked. This can lead to environmental movement activism in response to perceptions of democratic deficit on matters of environmental concern, as some social groups feel left out of the political arrangements based on corporatist relations. The closure on competitiveness by the state and its neo-corporatist partners in Ireland has also led to Irish environmental movements facing political closure. However, this exclusion also creates opportunities for movement challengers.

One result of the wider support enjoyed by environmental movements has been an increase in levels of professionalism and bureaucratic control. Within the role of NSMs themselves the role of movement leadership has evolved, as the requirement for expertise in areas such as management, media, politics and science has increased. The representation of environmental interests has also become dependent on a group's ability to translate an environmental issue in a manner that attracts the public's attention. State and corporate interests in environmental issues are being represented with a greater degree of sophistication due to public interest in environmental issues, and must be matched by increasingly sophisticated challengers. This may lead to a movement losing touch with more radical groups, and could create problems for challengers, as the authorities may exploit internal movement tensions, making movement success more difficult.

Local responses to a national or international environmental issue may vary in line with the availability of expertise within movement organisations. Additional factors which environmental movements must contend with include the extent to which control over environmental or development issues is centralised, the competency of the tier of government which must be dealt with, and the manner in which policies which affect local environments or communities are implemented (Carmin, 2003; Van der Heijden, 1997).

2.3. IRISH ENVIRONMENTAL MOVEMENTS

Perspectives on the future direction of the Irish environmental movement began to emerge during the years of growth synonymous with the 'Celtic Tiger' economy. As the contexts of environmental disputes shifted from concerns about toxic industries towards a debate about the problems of consumerism and waste in Irish society, the focus of environmental campaigns were redirected, while a debate commenced as to the future of environmental movements in Ireland. One of the key points of this debate centred on the need for movements to create stronger networks, in order to build an overall environmental movement in Ireland. The specific nature of localised or NIMBYist environmental campaigns in Ireland led to the creation of 'a particular kind of tunnel vision' (Cox, 1997, p. 33) for Irish environmental movements. The opportunities for bridge-building and movement cooperation were not to be seized by the disparate groups campaigning against pollution in their local communities. One NUI Galway ecology conference set out the 'aims, aspirations and realities' facing the Irish environmental movement. These included the establishment of 'a sustainable movement culture' (Cox, 1997, pp. 33–48), based on the movement theorists Mario Diani's definition of movements as 'networks between individuals' (*ibid.*).

This networking was defined as occurring between people involved in groups, rather than through groups themselves. Such a definition is similar to that of the role of 'interpersonal contacts' established by Friedman and McAdam (1992, p. 158), or of the 'solidarity networks' established by Della Porta and Diani (1999, p. 8).

The Galway conference, which was held under the title 'The Future of the Irish Environmental Movement', went on to discuss the significance of 'networks and cooperation' for environmental movements. The potential and possibilities of networks is based on two things; 'a shared movement identity' and 'a shared movement's culture' (Cox, 1997, pp. 33–48). Movement 'identity' is established through shared concerns and cooperative behaviour on environmental projects. 'Movement culture' is built from established contacts 'built into' (*ibid.*) specific campaigns, as well as into the lives of the participants. Within this understanding of networks, the concept of 'skills and understanding' (*ibid.*) is established. Cooperation networks can put activists in touch with those who have 'skills' needed by a movement. These individuals can be 'persuaded to come along and give us the benefit of their experience' (*ibid.*). This sharing of 'skills and understanding' (*ibid.*) could be used to build a more extensive movement.

However, the conference concluded that such networks had yet to be established in Ireland:

> I'm describing things, not as they are, but as they should be. We should be in this situation, but very often we're not-we don't see its value, we don't see what it's worth, we aren't prepared to 'waste time' on talking to each other except when we're already organising something very specific. And I think we pay the price of that. (Cox, 1997, p. 35)

The NIMBYist nature of environmental movements in the pre-Celtic Tiger years was perceived as a weakness that prevented the growth of a strong overall movement in Ireland. While the Green party had made small advances over time, they lacked the political strength or the links with the grass roots necessary to form a representative movement (Leonard, 1999, p. 66). The Greens had more success during European elections, and had moved away from the grass roots, signalling 'a commitment to electoral politics' (Mullally, 1999, p. 167). However, environmental movements remained 'locally based and relatively small' (Yearley, 1995, p. 653) groups of local residents who formed to combat the perceived threat to their environment (Baker, 1990; Tovey, 1992; Yearley, 1995) in the decades of pre-Celtic Tiger multinational-led industrialisation.

The lack of communication between groups remained a problem for Irish environmental movements throughout this time. The isolation experienced by environmental activists led to a lack of 'sustainability' in campaign, as 'burnout' became a feature of environmental activism (Cox, 1997, p. 36). Further difficulties were encountered in relation to 'motivation and mobilisation' (*ibid.*, p. 37) as activists struggled over time to maintain the organisation aspects of campaign mobilisation; maintaining offices, producing an 'alternative press' (*ibid.*, p. 39) and retaining the networks necessary for wider campaign mobilisation. This is linked to the creation and maintenance of a 'movement culture' (*ibid.*) which is dependent on a mobilising 'culture' (*ibid.*), which at that point, did not exist, creating 'the weakness of movement culture' (*ibid.*, p. 38). This weakness was evident in the 'terrible lack of strategic thinking' (*ibid.*) by the various elements engaged in environmental campaigns in Ireland. Further weaknesses were created through the importance of 'German ('fundis' and 'realos') or American (deep greens and light greens)' (*ibid.*) arguments about the contexts of environmental politics. Ultimately, this had led to a 'failure to connect' (*ibid.*, p. 40), as 'subdivisions' (*ibid.*) persisted in the various aspects of Irish environmental campaigning. This caused a 'failure to connect' (*ibid.*) within the various elements of Irish environmentalism, as 'movement identity' (*ibid.*) was not achieved. In the years after the 'Celtic Tiger' boom, GSE were able to overcome some of these obstacles due to their successful utilisation of communication technologies and mass media. This topic is dealt with in Chapter 5.

One recent study on the 'patterns of Irish environmental activism' (Garavan, 2004) has examined the manner in which one newspaper, *The Irish Times*, has reported on Irish environmental activism over the years from 1988 to 2002. This study also focuses on 'procedural forms of action' (Garavan, 2004, p. 13), taking a similar perspective as Taylor (2001). This study also shares Cox's (1997) concern about a perceived lack of movement coordination. One of the findings in the study contends that 'protest-orientated networking by Irish environmental groups is very low' (Garavan, 2004, p. 56), while the study's analyses of press coverage in the *Irish Times* finds that 'the reported incidence of joint protest action by Irish environmental actors is very low' (*ibid.*, p. 57).

The study also has concerns about the local or NIMBYist aspect of Irish environmental protest, arguing that its

> Data tends to support a characterisation of much Irish environmental protests as 'local'. (*ibid.*, p. 69)

Movement identification and mobilisation is also characterised as having a 'local' focus (*ibid.*, p. 92). The study uses *Irish Times* coverage to justify findings that argue that networking amongst Irish groups is 'very low' (*ibid.*, p. 112). The study concludes that most Irish environmental protest 'employs moderate forms of action' (*ibid.*, p. 122).

What is unique about GSE is that it formed networks at an early stage with other anti-incinerator campaigns, including those in Cork and Meath. The first was the Wexford branch of The Irish Farmer's Association (IFA), and the other was The Irish Doctor's Environmental Association (IDEA). Two statements were issued from the groups. The first was from the Wexford IFA:

> Farmers have responsibility for their produce ... emission from an incinerator ... could cause food quality concerns ... Therefore, Wexford IFA would urge Wexford Co. Council to reject the South East Regional Authorities Waste Strategy. (Wexford IFA December, 1999)

> The Irish Doctor's Environmental Association strongly opposes the plans for a waste incinerator ... our concerns relate to the adverse health affects from toxic substances in the emissions ... possible ill-effects include altered immune responses and disordered endocrine effects such as decreased fertility, lowered IQ and cancer. (IDEA, November 1999)

2.4. THE NEW MIDDLE-CLASS PROFESSIONALS: MOBILISING INTERNAL RESOURCES

'Resource Mobilisation Theory' (RMT) was initially conceived by McCarthy and Zald (1977) and later developed by many others. McCarthy and Zald revisited the concept of resources, extending its definition to include external linkages such as access to the state, corporate donations and other financial resources, as well as internal movement features such as expertise, professional networks and voluntary support from the public. Movements were seen to be dependent on externalities such as

> Societal support or constraint ... the linkages of social movements to other groups, the dependence of movements on external support ... and the tactics used by authorities to control and incorporate movements. (McCarthy & Zald, 1977, p. 1212)

Other studies of RMT (McAdam, 1982; Tarrow, 1994) began to look beyond material resources towards the importance of movement leadership, organisation and interaction with the political process. The response of movements to the political sphere can be measured by examining the strategic choices of its leadership, and the degree of organisation that can be achieved. Movement leaders must interpret the social and political processes that a movement encounters. As participation, commitment and achievement ebbs and flows throughout the duration of a protest, movement leaders or entrepreneurs must map out a strategic path, which can create mutual understanding out of shared grievances. The background of a movement's leadership is also a crucial feature of campaigns of collective action.

Whereas 'old social movement activists' were often comprised of members of the working class, NSMs (such as the environmental movement) draw their participants from what has been described as 'the new middle class' (Inglehart, 1977). Middle-class interest groups may mobilise around environmental disputes in order to increase their political inclusion in wider neo-corporatist arrangements, while at the same time defending their local communities' autonomy from the state's development of infrastructural projects in their locality. Environmental disputes may be exploited to create the leverage necessary to increase access in a system of political closure (Scott, 1990, p. 135).

Furthermore, the tendency in neo-corporatist systems to focus on 'technocratic criteria' (Offe, 1987 in Scott, 1990, p. 142), creates difficulties for political parties who wish to represent the concerns of the professional middle class, due to the tendency of professionals to have concerns which go beyond the economic, embracing lifestyle concerns as well. This void can be filled through the campaigns of environmental movements, as public sector professionals such as academics and those with alternative forms of expertise mobilise and challenge the structures of closed corporatist power. The critique of closed political systems applied to European states such as Sweden, Germany and Austria can be applied to Ireland's own system of neo-corporatist closure, echoing Scott's question:

> What is it about neo-corporatist arrangements that has stimulated the development of Green movements in those countries? (Scott, 1990, p. 144)

New middle-class professionals also tend to live in newly built, suburbanised areas, where urban sprawl can be seen to impact upon the surrounding hinterland. New middle-class groups may use their education and professional expertise to organise in response to their lack of political representation, together with local environmental concerns (*ibid.*). This combination of perceptions of democratic deficit and environmental grievance provides the momentum for movement formation as communities of professionals mobilise in defence of ecological responsibility campaigns. As white-collar public sector workers, these professionals are not as dependent on industrial growth as private sector professionals. They can oppose industrial development without facing the costs of such opposition. In this way, some new professional or middle-class members can afford to take on challenges to industrial policy in ways not open to other social groups included in corporatist arrangements.

Middle-class participants provide the leadership for movements. Zald and McCarthy (1987, p. 374) has described movement leaders as 'program professionals'. Leadership is exercised through the dissemination of information to potential activists through communication technologies. This allows 'program professionals' to set the ideological agenda of a movement at its inception, creating a type of leadership which differs from the formal settings of established political parties. Middle-class participants contribute to a cause by offering their time or expertise. Networks of potential supporters or activists, some with prior experience, are then activated. A rather different suggestion is that leadership elites may centralise decision-making power, and organise movements in a bureaucratic way, in order to increase movement effectiveness. These bureaucratic structures provide the basis for the maintenance of campaigns of collective action (*ibid.*).

The utilisation of various forms of protest is considered by RM theorists such as Gamson (1975) and McAdam (1982) to be an essential tactical resource. Due to their tactical importance, protests are planned and co-ordinated well in advance, with movement leaders orchestrating and demonstrating with the help of communication technologies. In so doing, a concerted attempt in made to highlight the movement's grievances, while attracting the support of the public to the movement's cause. In many cases, protests will embark on a series of disruptive activities that will then provoke a response from the authorities. In an era of continuous news coverage, contentious events are given priority airtime, and thus can capture the public's attention.

This mobilisation of wider support by a social movement has been called the mobilisation of 'solidarity networks' (Della Porta & Diani, 1999, p. 8). RM theorists define such solidarity networks as comprising like-minded individuals or groups who maintain association with movement leaderships through internet links and attendance at public meetings. In this way, RM theorists see potential movement adherents as part of a wider pool of professional and educated activists. This perspective maintains that potential activists are more integrated than in previous movements, as part of a wider solidarity network.

By defining movement mobilisation as a rational activity, RM theorists highlight existing networks of activists as a further resource. The alliances and networks formed by movements are an integral aspect of collective responses. Much of what transpires as movement activity has its roots in the activation of networks or 'interpersonal contacts' (Friedman & McAdam, 1992, p. 158). Movements are influenced by previous campaigns of action, and leaders may be drawn from prior movements. Networks may also be created from the sharing of resources such as office space and communication technologies. Networks provide outlets for participants from previous campaigns and the passing on of professional or scientific expertise from previous campaigns can provide crucial material for new challenges in other areas.

Once a movement's framework has been established, the interpretation of the issues in contention becomes a feature of a movement's organisation. Meanwhile, 'the mobilisation of consensus' throughout networks becomes central to the success of the movement at this stage (Klandermans, 1988, p. 182). As a sense of belonging is generated through the creation of a sense of shared grievance, the potential for the mobilisation of activists increases, and support for a movement's perspective is established through links with sections of the population with similar grievances, a movement begins to emerge. Once a congruent set of values has been established and spread, movement consensus becomes established. A movement must then be organised in such a way that this degree of mobilisation can be maintained. Strategic approaches can be introduced with activities such as:

> Public speeches, letters of opposition or support ... displayed communications ... pamphlets ... radio, television, audiovisual presentations, public events featuring well known intellectuals and political figures ... demonstrations ... door-to-door canvassing. (cited by Klandermans, 1988, p. 184)

Movements must maintain an array of tactical approaches. Tactics such as strikes or public marches present a challenge to authority, while also facilitating the promotion of a group's cause. This form of tactical approach also allows for a greater degree of interaction with the wider public, enhancing mobilisation while improving participant morale. Authorities are also notified of a movement's intent, and must respond in some way, either granting access or increasing repression. These 'repertoires of contention' (Crossley, 2002; Tilly, 1977) are the chosen tactics and strategies of a movement, usually passed on by movement leaders with prior experience. The history of past campaigns is an important feature in the creation of future repertoires of action. In this way 'collective action usually takes well defined forms already familiar to the participants' (Tilly, 1977, p. 143). As repertoires convey the medium of protest, media coverage has also come to shape the tactical choices made by movements. Large-scale direct actions, such as mass marches, are more likely to attract media attention. The increased media coverage of such events has led to a requirement of tactical flexibility and innovation from movement leaders. Fears of setting off reactions that increase repression may also influence repertoire choice.

Outcomes can be measured by the degree of structural or institutional change that result from a movement's campaign of action. Increased access to the structures of power is often synonymous with movement success, while increased repression may represent defeat. However, this is not always the case, as increased access may lead to a movement being co-opted by the political establishment, while increased repression may spark more widespread activism in response. By availing of the existing POS, a movement may achieve some degree of success, and so bring about change in the short term. However, movements that cross the threshold of the mainstream, and comes to be seen as political insider (*ibid.*) runs the risk of losing touch with its membership, or of being subsumed by the formal structures of politics.

2.5. THE EXTERNAL RESOURCES OF POS

There are limitations as well as the benefits of POSs that are relevant to the Galway case which I will discuss later. As societies undergo periods of transformation, increased discord can create what theorists have come to describe as 'political opportunity structures' that facilitate movement responses. This concept, first put forward by Eisinger (1973), describes the contexts in which patterns of political opportunities emerge and decline. The 'biases' of the political landscape must therefore be negotiated. If the political structure displays a degree of responsiveness towards activism, the opportunity structure for such groups remains open. Influence can be brought to bear by groups in this case. In the case of an over-centralisation of power, opportunities may be limited, as local interests are superseded by policy considerations.

The structure of political opportunities can therefore be derived from a number of indicators, such as the disposition of the political establishment to the concerns of protestors, the availability of and distribution of resources, and the existence of prior political expertise amongst a group's leadership. A group's ability to recognise the opportunities that surface during a campaign becomes a crucial feature of their campaign. In this way, the leadership of such groups become critical interpreters of potential opportunities. Political opportunities may be seen as a resource for movements, creating the societal strain necessary for collective action responses.

Further theoretical understandings of the POS have been put forward by Tarrow (1994). Opportunities may emerge from institutional reform after periods of transformation. Within this context, structural weaknesses in government may result in opportunities becoming apparent to groups or movements, who may then act, depending on circumstances whereby 'people gain the resources to escape their habitual passivity and find opportunities to use them' (Tarrow, 1994, p. 81).

Once activated, movements create further conditions for opportunities to surface, in response to mobilisation, emergent networks and responsiveness of government. If participation is widespread and effective, to the point where a successful outcome is possible, political opportunities may result. In addition, repressive or undemocratic responses by the state may create further opportunities for movements to exploit. Those participants 'who seize upon such opportunities have been described as movement "entrepreneurs"' (McCarthy & Zald, 1977 cited in Tarrow, 1994, p. 15). These entrepreneurs or leaders have the motivation to invoke activism through the utilisation of existing political opportunities. Tarrow defines POS as the external resources that can provide wider incentives for protests. The external features of a POS include aspects of the political system that affect the public, through government policies. These are enduring but informal with an uneven distribution, yet are utilised by every level of group organisation. The varied levels of political interaction provide the dimensions of opportunities which frame the political landscape from which a movement can emerge.

> The concept of political opportunity structure helps us to understand why movements sometimes gain surprising, but temporary, leverage against elites and then, quickly lose them despite their best efforts. (Tarrow, 1994, p. 85)

The changing nature of opportunity structures creates the varied circumstances for opportunity utilisation. Tarrow (1994, pp. 85, 87) puts forward four of the 'most salient changes' which occur in opportunity structures as 'increased access', 'unstable alignments', 'influential allies' and 'divided elites'. The first of these, the opening up of access, provides groups with the incentive to undertake a challenge. In this shifting political space, opportunities may arise. Movements that can maintain contact with adversaries through access may measure the effectiveness of their campaign, and are able to gauge the extent to which further actions may be tolerated. However, Tarrow also identifies an inherent risk which may result from prolonged access to authorities, as movement leadership may be 'cut off from their base' (Tarrow, 1994, p. 87), thus creating a loss of momentum in campaigns.

Another feature of the POS that can be exploited is the existence of political instability, resulting from electoral difficulties, uneasy coalitions and political realignments. The existence of political divisions in the ruling elite encourages challengers to act upon these 'windows of opportunity'. Electoral success in marginal constituencies and internal political disputes provide instances where opportunities can be seized upon by challengers. One way in which challenges can be undertaken is demonstrated in the third aspect of opportunity structure presented by Tarrow, the availability of influential allies. Gamson is cited in this regard, providing evidence which points to the existence of

> A strong correlation between the presence of influential allies and movement success. The allies can provide a legal or political counterbalance to the authorities, and provide advice or expertise to a campaign. The creation of interpersonal relationships and movement networks has become a feature of challenges to political elites, as the existence of such networks may provide increased chances of successful outcomes. (Gamson, 1990 from Tarrow, 1994, p. 88)

The existence of cleavages within elite groupings may present further political opportunities. Where divisions occur among those in authority, challenges can be encouraged. These political circumstances can provide the incentive necessary to induce mobilisation for action. Leaderships in waiting can also try to seize power by exploiting political divisions in the ruling establishment. Ultimately, these aspects of POS may not occur simultaneously, and the existence of any one of these features may be enough to encourage political challenges. Within this political upheaval, opportunities for protest may be found.

2.6. FRAMING THE EXTERNAL RESOURCES OF POS

Therefore, community-based social movements must exploit the political opportunities that emerge during the evolution of a dispute. One way that movements can successfully contest issues is by creating a degree of consensus around the meaning and values surrounding the dispute. By creating meanings based on their positions, movements can shape understandings about the nature of dispute, thereby influencing responses of the public and the involved institutions. This creation of meaning around protest is known as 'framing' (Snow & Benford, 1992). Movements create a frame by 'identifying events' (Goffman, 1974, p. 21), and establishing the sense of grievance that can be presented to the public as part of a movement's cause. The creation of meanings through 'collective action frames' is achieved in the following manner:

1. Amplification of meanings
2. Transformation of old meanings
3. Generation of new meanings (Snow, Burke, Warden, & Benford, 1986, in Snow & Benford, 1992, p. 136).

When a movement has identified a societal grievance which can provide the focus of a dispute, its meaning is 'amplified' by the mobilisation of resources, by that movement. This mobilisation process 'transforms' old meanings. Furthermore, it allows the movement to challenge the relevant institution or policy that causing the grievance. New meanings are then 'generated' through the 'cycles of protest' (Snow & Benford, 1992, p. 141) that emerged from that movement's ability to exploit the political opportunities that rise as a result of their challenge. This creates a link between the manner in which movements frame grievances, mobilise resources, and exploit political opportunities. This link is established around 'the mobilisation of consensus' (Klandermans, 1988, p. 175) as movements attempt to promote their perspective on issues through public events, utilising the media and by elevating their campaigns. By 'amplifying' certain issues of contention, movements can influence the cycles of protest, framing the wider understandings of these issues in a way that correlates with their ideological perspective.

According to Tarrow (1994, p. 123), 'frames like injustice are powerful mobilising resources'. Movement leaders, made up of entrepreneurs, must translate the grievances inherent in any perceived injustice to their movement, the public, and the institution being challenged. Movement leaders that fail to establish overall understanding about such grievances may lose control of the movements, and protests may lose direction as a result. One method of establishing consensus is through 'media framing', where movements strategically interact with the media, as the media becomes 'an external resource' (Tarrow, 1994, p. 126). This use of media allows movements to contest or create orthodox understandings about the issue or grievance central to the dispute. Internet technology and news networks can be utilised by movements to propagate their grievances and aims as a 'diffuse vehicle for consensus formation' and to 'help gain initial attention and maintain support' for their campaigns (Tarrow, 1994, p. 126).

Framing the significant events of a campaign can allow grass-roots movements to mobilise the data and expertise with which challenges to mainstream information sources can be taken. For environmental movements, it is their ability to use 'interest driven science' to challenge 'official science' (Grove-White, 1993, p. 22) that strengthens their case, and provides the leverage necessary to maintain campaigns. In this way, environmental movements have been able to frame wider understandings of what environmental disputes are about; allowing movements to gain increased access to POSs though a successful mobilisation of consensus about grievances, making it difficult for an institution or state to win the confidence of the public, who have become wary of risk technologies.

> The persistent political misrepresentation of human experience ... intensify mistrust in engaged individuals in environmental and other NGO's ... this results directly from the fact that the concepts ... are frequently inadequate ... people use, and even manipulate the consequential mistrust ... much of the drive behind environmental politics ... may well have arisen from it. (Grove-White, 1993, p. 25)

A global network of environmental and anti-toxics movements has emerged, providing resources, expertise and support for like-minded movements worldwide challenging 'the basis of social authority of scientific knowledge ... in environmental argument' (*ibid.*). By contesting official science, movements can frame 'a new moral discourse' (*ibid.*) around environmental orthodoxy, creating orthodox understandings that can become embedded in social thought. This allows movements to shape understandings, create consensus and to exploit the anxieties of the public, in an era where the 'risk society' has created simultaneous dependency and concern about technology (Beck, 1992). This has led to the 'opportunistic character' (Grove-White, 1993, p. 27) of much of the environmental movement's contestation of environmental policy. Political opportunities must be seized upon and framed in a manner that correlates with the agenda of populist movements. The shifting and evolving dynamic of the institutional dependence and need for regulation of technologies can lead to policy changes which are in turn exploited by environmental movements. However, the evolving POS that emerges around the contestation of orthodox understandings of environmental risks may have as much to do with the dynamic of movement agenda and institutional responses.

Some features or variables of a POS may act as both a resource for and a constraint on environmental movements. Recent literature, (McAdam, McCarthy, & Zald, 1996; Rootes, 1997a, 1997b) have built upon understandings created by Eisinger and Tarrow, on how POSs are established through the alliances and interactions which are formed by movement integration with the institutions of the formal political sphere. This external political environment contains a shifting array of variables from which collective activity is shaped. These variables include the structures of the state and society, prevalent ideology of government, and form of government response to challengers (Eisinger, 1973, p. 11). Movement entrepreneurs must create opportunities from these variables, thus creating a linkage between movements and wider political structures. Both Government legislators and movement entrepreneurs then translate the meanings that surround an issue. As government agencies promote policy implementation, movements contest the basis of a policy, challenging the state through movement activism.

However, these variables may not exist or may become problematic for interest groups in states that are characterised by tendencies towards centralised power, clientelism and corporatism. These are variables of the political environment that become the external resources that shape the political opportunities which movements can exploit. For instance, when the nature of government is weakened by coalitions, politics, internal competition, and corporatist arrangements, the resultant over-centralisation of power leads to reduced or closed political opportunities. Neo-corporatist structures may lead to the exclusion of environmental interests from partnership arrangements that focus on industrial and economic growth.

2.7. THE LIMITATIONS OF ACCESS TO POS IN A NEO-CORPORATIST AND CLIENTELIST POLITICAL SYSTEM

Political Representation through the Single Transferable Vote (PR-STV) allows constituents to maximise their voting power, and has led to an over emphasis on localism and personalism in the Irish system, as competition within parties between candidates becomes prevalent. Collins and O'Shea (2003, pp. 88–107) examine the manner in which a clientelist system 'facilitates rights and favours'. They see the clientelist system as a process of distributing goods and services to various sections of society.

> Brokerage work affects the operation of the political and administrative systems, and some suggest that it plays a part in shaping political culture. (Coakley & Gallagher, 1999, p. 225)

The Irish political system is presented as a clientelist model where corporatist structures incorporate pressure groups to the benefit of 'privileged interests' (Collins & O'Shea, 2003, p. 88). Collins and O'Shea also define Irish clientelism as a form of brokerage, liking it to a form of professional mediation, albeit with special access to the mechanism of state bureaucracy. This restriction of access may have implications for environmental movements that challenge policy. Environmental movements are dependent on a degree of openness in the existing opportunity structure to achieve the alliances necessary to further their representation of interests. Movements that engage in challenges to policy are dependent on the availability of resources and opportunities within existing political environment. Problems arise for movements when collective behaviour encounters restricted access to the levels of power, due to the prevalence of clientelist structures or over-centralised institutions of government. Political closure through the centralisation of power can result in weak governments, and an over reliance on coalitions. The distinctive types of institutions that shape the wider political structure are central to understandings of the performance of environmental movement operating within a given political framework (Scruggs, 1999, p. 9). In the case of governments who favour corporatist arrangements, environmental movements may face restructured access. If corporatist policy-making excludes environmental groups and prioritises the interests of industry, this restriction may constitute an opportunity for movements who wish to challenge the resultant policy. As professional middle classes supply a large number of environmental movements' participants, the exclusion of this social group's concerns from the corporatist table creates unrest and gives rise to movement responses (*ibid.*).

In regimes such as Ireland, that use a system of PR-STV, a wider representation of interests may occur, as party competitions compels 'parties to accommodate interests' (Scruggs, 1999, p. 9). Local councillors keep the brokerage system activated while the TD is away on Dáil business. Bax (1976, p. 81) calls these local brokers 'the broker's broker'. While this set of political relations promotes alliances between interest groups and politicians at a local level, the demands of government from the centralised core leads to a restriction in opportunities, as access is constrained. However, in states where clientelism is prevalent, this may not be the case, as competition between the parties may overcome such accommodation of interests. As movements seek out alliances with political parties or figures at times of elections, promises given in return for electoral support may be broken, as the demands of government formation supersede all other considerations. In this way, the political opportunities presented to a movement can be demonstrated to be constantly in a state of flux, highlighting the importance of a movement's ability to negotiate the shifting sands of the POS open at the moment collective action is undertaken.

2.8. STATE POLICY AND NEO-CORPORATISM

The priority of the Irish state since its inception has been the creation of a stable economic environment that would provide a platform for long-term employment and continued economic growth. The ideological and electoral concerns of the political parties have been overcome by the prioritisation of low corporate tax regimes, beginning with the Lemass/Whitaker Shannon Enterprise project in the late 1950s.

The focus on low corporation taxes was one of the policy goals of the Fianna Fail/PD coalition of 1992. It was made a priority of the semi-state body the Irish Development Authority throughout the 1970s and 1980s, and was one of the policy goals of the Fianna Fail/PD coalition of 1992. The decision to focus on neo-corporatist partnerships was a strategic choice for Fianna Fail, as that party had a populist core identity. Fianna Fail had made a choice to favour the state's policy framework in order to maintain focus on competition and growth. For instance, in regard to the 2002 election, Fianna Fail made no change to their waste manifesto, despite populist pressure. According to Bartolini and Mair (1990 from Mair, 2001 p. 24), relationships between ideologically opposed parties, which are presented in manifestos for election, may differ from the policy pursued by that party once elected. In other words, the ideological contestation of elections may have no bearing on the subsequent policy programme of that party once elected. Invariably, Fianna Fail reverted to the state's competitive policy framework once they had been re-elected. The merger of party manifesto and state policy created an overarching focus on a 'multifaceted competitiveness' (Mair in Laver, 2001, pp. 23–24).

Today Irish policy remains in the hands of state managers, technocrats and un-elected bureaucrats, advisors and civil servants. These elites are recruited by the state to maintain its structuralist supremacy over the ever-changing political actors who are concerned with populist responses to their stated manifestos. It is the job of the technocrat to ensure that the state's political strategy of maintaining competitiveness and growth is maintained, in spite of populist or ideological demands. In this way, environmental challengers are stymied, as their local political allies are restricted in their responses once they have entered parliament.

The state has reached out to the major representative social groups such as the business and trade unions to create a neo-corporatist social partnership that shared this focus on competitiveness and growth. Here, an agenda for low corporate tax regimes, improved industrial relations and wage restraint was established as the template for competitiveness and growth. The concept of 'bargained' or 'neo-corporatism' (Roche and Cradden cited by Adshead & Millar, 2003, pp. 71, 74) has the state taking a pro-active lead in a collective bargaining process which encompassed employers and unions in an attempt to manage economic performance. This form of 'competitive corporatism' subsumes the once divergent agendas of tax reform or reduction, flexible specialisation in the workplace, pay negotiations, industrial relations, training of the work-force, and reform of public and state expenditure. Leading employer and trade union representatives took on the role of issue groups' negotiators trying to gain concessions from the state for their members. Forms of neo-corporate social partnerships have been denounced for being undemocratic (*ibid.*) as this type of institutional arrangement may lead to an over-centralisation of power, with the prioritisation of focus on the concerns of social elites over the aspirations of the democratically elected parties of the Dáil.

Another perspective on social partnership is that it allows for an enhancement of democracy, as political participation becomes more inclusive of sectional interest groups (Hirst, 1993). However, the model of social partnership in operation in Ireland currently excludes environmental movement organisations, creating ongoing tension between the state and the environmental lobby. As environmental concerns may run counter to the state's prioritisation on competitiveness and growth, a dichotomy between both the state and the EC's environmental directives on one hand, and between the state and local environmental challengers on the other, has opened up, creating the opportunity for environmental challenges to state policy to be undertaken.

2.9. THE 'PARADOX' OF NEO-CORPORATISM FOR ENVIRONMENTAL MOVEMENTS

Critics of corporatist and neo-corporatist centralised arrangements (Kriesi & Tarrow, 1990; Scott, 1990) have indicated that the processes of inclusion or exclusion that result from the state's facilitation or repression of access to political structures creates a 'dimension of political opportunity' (Tarrow, 1994, 1998) for movements such as environmental campaigners. 'Inclusive' corporatist arrangements are usually 'restricted to employers and organised labour' (Scott, 1990, p. 144). The 'closed' (*ibid.*, p. 145) nature of corporatist arrangements:

> Means that groups excluded from these processes may mobilise at grassroots level, knowing that 'normal' challenges are closed off. (*ibid.*)

One social group that is particularly affected by this form of corporatist closure is the new middle-class professionals. This group is therefore in a position to challenge political exclusion or democratic deficit, providing new middle-class advocacy groups with a degree of oppositional power now lost to the now 'less significant' (*ibid.*) unionised industries.

> It is the paradoxical position of the new middle class, rather than its exclusion alone, which has inclined it towards social protest and ecological ideology. (*ibid.*)

This 'paradoxical position' has been described as a 'simultaneity' between the 'challenging' and 'sustaining' dimension of rights campaigns and institutionalisation, placing rights campaigns between 'the power over' and 'the power to' within the wider dimensions of POSs (Stammers, 1999, p. 996).

2.10. NIMBYISM, ECOPOPULISM AND DEMOCRATIC DEFICIT

The context for environmental movements' opposition to toxicity can be traced from the 1970s and 1980s in the United States. During this time, public opinion began to focus on the risks posed by toxic industries due to highly publicised incidents of pollution such as the contamination of the Hudson River in New York, the Love Canal toxic dumping controversy and the Three Mile Island incident. Such controversy gained large-scale media interest, and news networks began to cover toxic spills with greater emphasis (Szasz, 1994, pp. 38–48).

The issue of toxic pollution became a part of contemporary culture, as tanker spills and toxic clean-ups became commonplace on the nightly news. The public became aware of the fact that toxic pollution was an issue that could affect their locality, 'personalising' the toxicity issue (*ibid.*). Once the toxicity issue had been located in local community thought, grass-roots political responses to toxic pollution soon followed. Examining the process of how the issue changed from 'mass perception to mass action' (*ibid.*, p. 67) Szasz outlines the manner in which local concerns about toxic pollution led to the opposition dubbed 'NIMBY' or 'not in my backyard' groups.

From their NIMBY inception, anti-toxics groups in the United States began to share information, establishing a network infrastructure. Protests occurred across North America, and media coverage of these protests was extensive as the debates about toxics emerged as a political issue. But as the anti-toxics movement established, it gradually moved beyond NIMBY into what has been described as 'ecopopulism' (Szasz, 1994, p. 69). The realisation that NIMBY perspectives would only lead to duplicity or secret dumping led grass-roots groups to realise that a more organised movement needed to be maintained. Szasz refers to this progression as being one 'from NIMBYism to radical ecopopulism' (*ibid.*). The creation of a wider, focused concern about toxic pollution creates the type of ecopopulism necessary to maintain campaigns against pollution industries and the state.

In the Irish populist political context the state is unable to respond to the concerns of the population on an issue such as toxic risk, a form of 'democratic deficit' occurs. Democratic deficit can be compounded by the neo-corporatist considerations of the state and its industrial partners. It can also occur because of the state's reliance on un-elected 'technocratic' advisors (Held, 1987, pp. 178–185) as policy is formed on the basis of advice and expertise provided by technocrats. Although the Irish political system is populist, such populist concerns are superceded by the state's dependence on technocratic expertise. Seen in this light, the technocratic state becomes an 'elite' (*ibid.*), imposing policy options on communities despite any local opinion or concern. This elite approach to technocratic policy implementation has implications for local government, which may be closer to local concerns. It also has similar implications for representative politics, as local politicians who wish to raise such issues have their concerns overruled. By creating such a democratic deficit, the state creates political opportunities for ecopopulist movements.

2.11. THE US ENVIRONMENTAL JUSTICE MOVEMENTS: FRAMING ISSUES OF HEALTH, DEMOCRATIC DEFICIT AND NIMBYISM

Further understandings of the Irish case can be gained by examining issues which emerged around environmental justice in the United States. According to Szasz (1994) a distinction can be made between the localised impact of a toxic waste 'event' such as Love Canal and how it is then presented as a media issue at the national level. For Szasz, the media's role in getting an impact nationally from a local environmental issue is crucial to the success of that issue as a target of movement grievance. This impact can be measured by the airtime given to the Love Canal issue:

> The network's nightly news programs alone carried about 190 minutes of news about hazardous waste. (Szasz, 1994, p. 42)

This large amount of prime-time news coverage led to an overall shaping of the issue of toxic waste, into what came to be a societal understanding of what toxic waste problems are about in the United States. As such, local activism was picked up on by the national media, who then played a major part in regard to the building of consensus about toxic waste issues. One result of this increased coverage was an increase in localised activism. After Love Canal:

> Contamination protests that had started earlier but had floundered suddenly took off after 1978. (Szasz, 1994, p. 70)

A sense of momentum built up around the grass-roots anti-toxics movements, as people found a sense of inspiration from the success of that movement. By 1979 the Waste Disposal Industry was facing a 'major barrier' (*ibid.*) from public resistance to toxic dumping, whereas previous to the Love Canal campaign, the issue was not capturing much public attention. What resulted was a major siting problem for the toxic waste industry in the face of 'inflamed public opinion' (*ibid.*). The strength of the anti-toxics movement in the United States led to a support network which could facilitate fledgling movements in Canada, and Europe (including Ireland), and ultimately worldwide.

At the centre of the nationwide anti-toxics movement in the United States was the 'Citizens Clearinghouse for Hazardous Wastes', which in itself evolved from the Love Canal protests. The CCHW endeavoured to bring together regional groups that were embarking on campaigns, and provided them with 'start-up' information about conducting a protest. Furthermore, a National Toxics Campaign based on localised source reduction has been established. The NTC also targets toxic waste incinerators, and offers 'organising help, technical assistance and laboratory testing' (Szasz, 1994, p. 75). In this way the NTC has parallels with the CCHW, as both groups have become a major resource for anti-toxics campaigns in their early stages.

The utilisation of technology is maximised through the input of the academic sector, which provides information on the most recent scientific research into toxics. This information is broken down into a set of prioritised findings, and posted on newsletters such as the 'Remote Access Chemical Hazards Electronic Library (RACHAEL) Hazardous Waste News'. The utilisation of technology and expertise has facilitated anti-toxics movements across class and racial divides, in both national US and global settings. This has allowed anti-toxics movements to maintain an immediacy that formal political structures would hamper. Another benefit of the dissemination of anti-toxics information has been the altering of activist demographics in relation to class and racial profiles. Historically, toxic waste disposal facilities were sited in poor areas of high black, Latino or Native American populations. These areas were deliberately targeted as:

> Some policy analysts began to advocate a strategy of siting in communities that are least capable of politically resisting or more amenable to accepting some form of financial compensation in exchange for accepting the facility. (Szasz, 1994, p. 75)

2.11.1. Framing the Issues

As the anti-toxics movement in the United States grew as an issue, the larger environmental groups such as Greenpeace and Friends of the Earth became advocates of the cause. This allowed the anti-toxics issue to expand and become part of the overall environmental agenda, as toxic dumping was linked with broader issues such as ozone depletion. This expansion of the anti-toxics movement is also illustrated by the shift in the characterisation of such movements as having NIMBY or 'Not in my Backyard' perspective. The labelling of anti-toxic agitation as NIMBYist is seen by writers such as Szasz to be an attempt to reduce the impact of the professionalised links which have become a feature of these movements. The success of the US anti-toxics movement in 'inflaming' public opinion has invariably led to an increase in the exportation of toxic wastes to poorer states in the Southern Hemisphere.

A feature of the anti-toxics movement in the United States has been the anti-incinerator movement. After an initial problematic phase in the 1960s, the incinerator industry developed new semantics for their operations, such as 'waste to energy' (WTE). A new era of efficient and cleaner incinerators would convert electricity from the process of 'thermal treatment', under the eye of a more vigilant industrial and federal sector. However, a number of the new mass-burning plants experienced difficulties at an early stage. These included:

> Large volumes of ash, emission problems, excessive operating costs, and difficulties in marketing electricity. (Walsh, Warland, & Smith, 1997, p. 5)

Between 1980 and 1990, the WTE incineration industry grew, in spite of these difficulties. One major reason for this was the public's increased intolerance for any form of hazardous dumpsites, while municipal waste sites were becoming full. Throughout this time, the Reagan government favoured private disposal firms rather than encouraging waste reduction. This added an ideological aspect to the WTE versus waste reduction debate. Also by the mid-1980s, scientists such as Barry Commoner and Paul Connett began to highlight the health risks posed by the dioxins released by the process of incineration. These two advocates of the anti-incineration movement played a prominent role in challenging the orthodoxy of the WTE industry. Their impact on the issue can be understood through the following statements:

> Barry Commoner was the first to make dioxin out to be the most threatening human carcinogen that ever appeared ... Commoner was viewed by the public as the major guru on dioxin. (Carolyn Konheim WTE industry spokesperson 1989 from Walsh et al., 1997, p. 8)

> By the late 1980s, if an industry proponent were asked to identify a single individual symbolising the opposition, Connett's name would probably be the first one mentioned. (Walsh et al., 1997, p. 17)

The dissemination of information on the health risks of incineration by scientists such as Commoner and Connett led to an increase in mobilisation against the incineration industry. This professional level of expertise was combined with the ongoing technological difficulties the industry was facing, to create enough public interest and issue consensus to create an anti-incineration movement, centred in the North East of the United States, but with links to existing anti-toxics and environmental groups. The issue of the dioxin content of incinerator emissions became the prominent theme of this movement. In particular, chlorinated dioxins (PCDDs) and chlorinated furans (PCDFs) were cited as potential hazards (Denison & Ruston, 1990, p. 9).

The major issues involved in the creation of health risks, according to the protestors, were the heavy metals contained in emissions, and the difficulties with ash residue disposal. The main alternative to incineration put forward by the scientists was recycling. The emissions from incineration contained carcinogenic heavy metals such as furans and dioxins, and other metals such as arsenic and cadmium. The presence of dioxins in emission gases of municipal waste incinerators became one of the main issues for the anti-incinerator movement. The other major issue highlighted was the problem of ash residue, as the ash residue is both toxic and non-combustible. The issue of ash residue has become one of the biggest obstacles to the incineration industry's gaining acceptance with the wider public:

> Critics consider it the Achilles' heel of the modern incinerator industry because the more successful a given facility is in removing toxins from the air, the more toxic its residual ash necessarily becomes. (Walsh et al., 1997, p. 11)

According to US Congress Reports, both the fly ash, which becomes airborne, and the bottom ash, which is found under the incinerator's grate, contained toxic minerals. In addition the incinerator process does not reduce waste beyond 15 to 30 per cent weight and 5 to 15 per cent volume of the original solid waste material. Therefore, a substantial amount of waste material, now containing harmful toxins, is left over after the incineration process. The harmful residues contained in fly and bottom ash, come into contact with air, land and water supplies, according to US Federal sources:

> Substances may leach into ground water, and airborne or waterborne transport of ash during handling operation or from landfills may lead to inhalation, ingestion in food crops, or dermal exposure. (Office of Technology Assessment, US Congress, 1989, p. 254)

The main alternative put forward by the anti-incineration movement is recycling. Recycling also has a variety of meanings in different situations. In regards to it being an alternative to incineration, recycling involves the separation, reduction and reusing of solid industrial waste materials, and the composting of household waste. The effectiveness of recycling is contested between the incineration industry and the anti-toxic groups, with potential targets ranging between 'a low of 15 percent ... to a high of 90 percent' (Walsh et al., 1997, p. 12). Paul Connett is at the centre of the global 'Zero Waste' movement, which aims for a complete elimination of waste through a comprehensive programme of recycling. Recycling also presents an alternative to the landfilling of waste, which has increasingly become problematic due to the lack of available space. This has led to a situation whereby incinerator industry spokespersons cite incineration as the answer to the problems of landfill, while the anti-incinerator movement argues that waste reduction and recycling counters the need for landfill and incineration. Another major area of dissension for both groups is the high level of waste needed to supply the process of incineration, as opposed to the low costs of the recycling process. As such, the debate often takes on a global growth versus a localised constraint aspect, as the economic viability of incineration or recycling is assessed.

2.12. CONCLUSION: MOBILISING INTERNAL RESOURCES AND EXPLOITING EXTERNAL OPPORTUNITIES

This chapter has examined the literature that addresses the manner in which movements mobilise internal resources and exploit external opportunities while operating within a framework of political structures. The relevance of a wider POS for environmental movements that operate in a clientelistic and neo-corporatist state has been highlighted. As opportunities are presented, the role of movement leaders becomes important, as the crucial nature of movement mobilisation in response to attempts to gain access becomes apparent. The role of new middle-class movement leaders in challenging policy has been outlined.

In the first section, the emergence of NSMs has been detailed, placing environmental movements in the context of collective actors who challenge the state for reasons other than concerns about economic inequality. Environmental movements can be seen as a part of the NSM 'sector', displaying characteristics in their mobilisation and organisation that are typical of 'new' movement campaigns.

Environmental movement participants are often drawn from the post affluent 'new' middle class that has concerns about lifestyle issues, autonomy and preserving local hinterlands. This chapter has presented an understanding of the manner in which environmental movements organise, mobilise and exploit opportunities. Particular attention has been paid to the problem faced by environmental movements due to restricted access to closed or centralised power at national level, which supersedes local political alliances forged by movements and local political actors. States that are characterised by clientelist relations and neo-corporatist arrangements may or may not create restrictive POSs. While networks and alliances at local level may facilitate access, the demands of centralised power in closed systems may ultimately restrict access, leaving environmental movements outside of neo-corporatist partnerships that focus on industrial growth over other concerns.

The environmental concerns of the new middle class are ultimately seen to be contravening the economic priorities of the neo-corporatist partners of the state, business groups and trade unions, which are focused on maintaining competitiveness (Jessop, 1990, pp. 248–272). This chapter has examined the 'paradox' of neo-corporatist closure for environmental movements, as such political arrangements deny access on one hand, while also providing an impetus for the mobilisation of environmental campaigns on the other (Scott, 1990). While environmental movements may mobilise local populist support and forge alliances, the POSs that facilitate this at local level may lead to restriction of access to the political arrangements of core state structures. Ultimately, the manner in which movements mobilise internal resources such as expertise, networks and leaderships to frame their campaigns allows for an extension of these campaigns into a phase of exploiting the external resources of the POS.

These movements have come to be seen as 'rational actors, facilitated by the mobilisation of resources by movement leaders of entrepreneurs' (Della Porta & Diani, 2006 [1999], p. 7). Tilly (1978, pp. 52, 53) has established an elementary 'polity' model depicting the wider participants in many forms of collective behaviour responses to social discord. This includes the participation of 'government', or the institutional authority causing social grievance, 'contender' which contains the collective activity of groups and governments, and 'coalitions', which are networks that emerge from such collective action.

Tilly (1978, p. 55) also details a further component in the process of campaign emergence, which is 'opportunity'. POSs are established through the interaction of a movement with the wider political sphere. Tilly (*ibid.*)

describes this POSs as being comprised of the following elements; 'power' or the favouring of interactions to the benefit of one side of a dispute. 'Political power' results from 'interactions with governments'. The degree of power lost or gained is equated with the degree of 'favourability' achieved. 'Repression' is measured through the 'costs' of activism extracted from contenders. 'Political repression' is derived from the increased costs that result from losses after a challenge to government. 'Opportunity' can be seen as the degree of success of a groups or 'realisation of its interests' are being enhanced, through a successful utilisation of events and resources.

In his article on 'Identity and Mobilisation', Melucci (1988, p. 329) sets out an understanding of how the 'gap' between the objective conditions and 'collective behaviours' created the phenomenon of an 'actor without action'. Melucci (1988) has fashioned an understanding of the dynamics of group integration and mobilisation. Individuals communicate meanings, which are translated into action. According to Melucci, a process of 'organisation' takes place in the following manner:

> Individuals acting collectively 'construct' their action by means of 'organised' investment: that is, they define in cognitive terms the field of possibilities and limits which they perceive, while at the same time activating their relationships so as to give sense to their 'being together and to the goals they pursue. (Melucci, 1988, p. 332)

Melucci's concept of the group dynamic can be applied to Galway for a Safe Environment's (GSE) committee, who put aside potential difference to formulate their campaign. Furthermore, Melucci has highlighted some of the underlying aspects of group mobilisation, which can address the case. His understanding of 'movement potential', or the existence of a section of the population who are favourably disposed to a campaign, is applicable to the environmental campaigns included in this study. Mobilisation has been defined by Etzioni (1968, pp. 388, 389 cited in Tilly, 1978) as 'the process by which a unit gains significantly in the control of assets it previously did not control'.

Tilly (1978, p. 69) presents Etzioni's classification of assets of resources that are utilised for mobilisation purposes. These include 'coercive' and 'manipulative technologies'; utilisation, such as finances and 'information services' and/or the loyalties of interpersonal contacts. The resources mobilised include information and communication technologies, the information services of mainstream media, local and national press and the internet, and the normative networks and contacts with local and international figures who could provide expertise for the case. In this way, movements are able to interpret 'shared grievance' such as the proposal for an incinerator and shape a challenge to government waste policy from this interpretation. One of the first challenges faced by the committee was convincing the public of their argument. This process is described by Piven and Cloward (1979) as follows:

> It is also necessary that people who normally accept authority and are convinced of the legitimacy of institutions come to recognise that this authority and these institutions are unjust and wrong. In addition, people who are usually fatalistic and feel that the existing order cannot be changed must become convinced that they are capable of changing their condition. (Piven and Cloward cited by Klandermans, 1989, p. 179)

While the shared grievance of opposition to incinerators brought the participant together, there were those in the committee who recognised the need to avoid generating the degree of opposition faced by previous environmental groups in the media. Klandermans (1989, p. 181) has outlined the importance of establishing 'action goals' which are:

> Instrumental for eliminating the dissatisfaction or fulfilling the aspirations that are at the root of a movement's mobilisation potential. (Klandermans, 1989, p. 181)

Studies have indicated the existence of 'leadership types' who step forward as groups emerge. Turner and Killian (1987) set out an understanding of 'symbolic leaders' who come to represent the group. On a more practical level, 'decision-making' leaders 'direct the movement by their preferences and choices' (Klandermans, 1989, p. 220). Leaders can be further distinguished as mobilisers of consensus, publicity or other resources. Wilson (1973) identifies these distinct leadership types as being either 'charismatic' or 'pragmatic', while further studies indicate a distinction between participants who engage in bureaucratic organisation, and other who focus on activism. The utilisation of resources by leaders shapes the structures of a movement. In GSE's case, the creation of a sense of grievance about the potential health risks of incineration was an indication of how its committee were successful as leaders. Later disagreements about alliances with political figures demonstrate the pitfalls of a committee approach to leadership.

A central feature of this analysis is the importance of how issues are framed in the context of social grievances. The relevant issues can only be framed through the intervention of movement entrepreneurs or gatekeepers. 'Issue importance', according to Dearing and Rogers, can be increased through 'public' and 'policy agenda setting'. From this perspective, policy response to an issue is dependent on the importance such an issue is granted by public and media agendas. In this way, successful outcomes to social movement activity is demonstrated to be increasingly reliant on that movement's ability to frame issues and influence professional media elites, in addition to provoking a public response to an issue of public concern. Subsequently, a movement's ability to frame issues and build consensus becomes a crucial aspect of their campaign.

A further understanding of the importance of influencing media in the framing process can be achieved through an analysis of the 'editorial gatekeeper'. It is the role of such gatekeepers to decide what events in society are noteworthy enough to feature on front page or prime-time news reports. Their importance, to a grass-roots movement framing their grievance, can be crucial to the success of a campaign. While political disputes still require the exchange of views from adversarial groups, the influence of media on the public perception of that dispute ultimately influences the outcome of such disputes. Gradually, the media's coverage of events can be influenced by ideological-driven perspectives. The meanings of societal disputes around topics such as environmental or human rights are subject to an ideologically based conflict of definition. Here again, framing issues through the media plays a central role, shaping the public perception of social issues.

According to Tarrow (1994, pp. 125–129) media framing is an important aspect of movement strategy. Tarrow looks at 'three stages' where media becomes 'an external resource' (*ibid.*) for movements. This first is in the area of consensus mobilisation and formation, from a movement's inception through the establishment of a shared social grievance that can be framed as an injustice. Secondly, the media allows movements to communicate with the public, providing movements with a degree of legitimacy and 'status' (*ibid.*). However, Tarrow also argues that the media is 'far from neutral' (*ibid.*) and movements must establish their credentials through the provision of accurate data to help the media's presentation of the issue. In so doing, movements can give a degree of journalistic or editorial support. Movements have become adept at utilising internet technologies as a resource for communicating with the mass media.

CHAPTER 3

THE EVOLUTION OF POS: ECONOMIC GROWTH AND THE MOBILISATION AND FRAMING OF COMPARATIVE CASES

3.1. INTRODUCTION – THE TWO PHASES OF MULTINATIONAL-LED ECONOMIC DEVELOPMENT, EMERGENT CONSUMERISM AND COMMUNITY RESPONSES: 1958–2005

Environmental campaigns in Ireland can be divided into pre- and post-affluence phases of multinational-led economic development in Ireland between 1958 and 2002. I construct an 'issue history' (Szasz, 1994) that locates the GSE case within the context of a series of community challenges to toxic industries during that period. These cases are relevant to an understanding of environmental movement activity in the decades before GSE's own campaign, before the onset of economic growth, or in the post-boom years when issues such as landfilling and incineration were the subjects of disputes across Ireland. I examine some earlier cases that have parallels with GSE's case in terms of their primary focus on toxicity and health risks, but I also look at the emergence of landfill and municipal incineration disputes since the mid-1990s that can be traced to the explosion of waste as a by-product of Irish society's fixation with consumerism.

The cases are also examined through an application of Tarrow's 'four most salient changes.' The chapter applies Tarrow's 'four salient changes' in relation to the 'dimensions of political opportunity' to each of these community responses. These aspects of the political opportunity structure include 'increasing access', 'unstable alignments', 'influential allies', and 'divided elites' that can encourage community responses to state policy (Tarrow, 1994, pp. 85–89). The chapter uses Tarrow's theory to shed light on the following questions, with a summary of each case being provided at the end of the relevant section:

1. Who is looking for 'increased access' in these cases and how is access increased?
2. What types of 'alignments' were created?
3. To what extent were alliances 'influential' or otherwise and who were the influential allies?
4. How were elites divided?

The cases included in this chapter, as well as providing us with a range of comparative cases with GSE, also provide us with a frame work with which we can test and contrast the mobilisation and exploitation of internal resources and external political opportunities.

An issue history survey provides us with further understandings of the nuances of each case, placing them in the overall context of environmental campaigns in Ireland that have mobilised and framed issues concerning health risks and democratic deficit during an era of post modernity (Lyotard, 1984) where:

> Intellectual life is now dominated by a crisis of faith in the major modern ideas of science, progress and reason; individuals have lost their trust in the ability of science to give them answers. (Bilton et al., 2002)

This 'crisis in faith' was manifested in the community challenges to multinationals in Ireland over this period as the concept of 'science' itself came to be contested. Another significant event that shaped the 'curvilinear' (Eisinger, 1973) POS surrounding these issues was the emergence of the 'Peace Process' and the signing of the Good Friday Agreement in 1998. This led to increased opportunities for cooperation between the administrations in Dublin and Belfast, as the international funds for reconciliation turned the border into a resource, creating 'networks of interaction' (O'Dowd & Wilson, 1996, p. 8) in the once mutually exclusive areas on either side of the frontier.

The wider context of the evolving POS in this case was also influenced by external factors such as the demands placed on the state by EU anti-pollution directives. This increased demand for a more stringent regulatory framework, in addition to the increase in community campaigns on environmental issues, led to the creation of the EPA in 1992. The establishment of the EPA in conjunction with an overall regulatory framework for pollution control is explored in detail in the next chapter. However, the formation of the EPA is included as part of this chapter's structure, to illustrate the transition undertaken by the state, as it shifted from its primary concern with multinational led growth in the 1970s and 1980s to a somewhat contradictory concern with pollution control in the 1990s and 2000s.

Within the pre- and post-Celtic Tiger phases presented in this chapter, a series of sub-categories are established as part of the evolution of the issue history and POS. These sub-categories can be outlined as follows:

- Pre-Celtic Tiger disputes: opposing toxic industries
  1. Anti-multinationals: Raynestos Manhattan/Merrell Dow
  2. Anti-pharmaceuticals: Merck Sharp and Dohme
  3. Cross-border I: Du Pont
- Post-Celtic Tiger and EPA: regional waste management
  1. Anti-landfill: Tralee and Galway
  2. Anti-incinerator: Cork Harbour and Duleek, Co. Meath
  3. Cross-border II: All Ireland Waste Strategy

By placing these campaigns in this context, we are able to establish exactly how the political opportunity structure evolved in a 'curvilinear' (Eisinger, 1973) manner, as the shifting dimensions of POS came to be affected by external events such as the peace process or the onset of environmental regulation. This application of an 'issue history' approach to the relevant cases that preceded or ran in parallel to GSE's campaign provides a framework with which a contextualisation of environmental campaigning in Ireland can take place. The variables of resource mobilisation, framing and interaction with the POS are tested by applying Tarrow's (1994, 1998) 'four most salient changes' to each of the cases that have emerged in this context. In so doing, the chapter sheds light on whether the transition of environmental movements from their NIMBYist inception towards a radical environmental perspective, as defined by Szasz (1994) has been undertaken in any of these cases.

The POS was established through the state's choice of a policy that created a dependency on external trans-national investment at a time when groups of 'new middle class' professionals were becoming concerned about environmental issues. The state was also in the paradoxical position of having to implement EU environmental protection policies, which the local authorities were responsible for while continuing to compete with each other to attract multinationals into their locality (Coyle, 1994, p. 68). This 'vigorous pursuit of a policy of attracting foreign investment to Ireland' (Smyth & Boylan, 1991, p. 156) became the priority of both the state and local authorities during the 1970s and 1980s. At the same time,

Irish communities were beginning to develop 'from below' in a community led manner (Varley, Boylan, & Cuddy, 1991, p. 49) as groups opposed to the health risk potential of some multinationals were formed. At the heart of the matter was the prioritisation of economic growth over environmental protection, leading to 'an extremely facilitating and generous 'open-door' policy to foreign companies' (Smyth & Boylan, 1991, p. 156). Ultimately community campaigns against multinationals emerged from the political opportunities thrown up by this paradox at the same time as the potential for mobilising the concerns and expertise of a middle class that had grown by 22% in the previous two decades (Breen, Hannon, Rottman, & Whelan, 1990, p. 57) was increasing.

The chapter also clarifies the two strands of 'scientific data' utilised by both communities and the state in the course of disputes. On the one hand, a form of 'official science' (Grove-White, 1993) exists, based on the data of the professional consultants such as chemists and engineers employed by the state. This 'official science' is based on understandings of science and technology that are linked to industrial development. Then there is the 'issue driven science' (*ibid.*) utilised by communities who challenge state policy based on concerns about the health-risks posed by the processes of 'official science'. Professionals such as Dr. Paul Connett contributed the 'issue driven' scientific evidence in the DuPont case in Derry and in toxic disputes in Cork Harbour, as well as being a pivotal resource in the GSE case.

Disputes surrounding three industrial plants, Merck Sharp and Dohme, DuPont, and Raybestos Manhattan, are explored in this context of the background to the emergence of 'interest driven' scientific responses as part of the framing of Irish environmental campaigns. Each of these was a siting dispute where local communities mobilised around concerns about toxicity and health-risks. They each have parallels with GSE's case, as local communities contested the official scientific data of the state and corporate sector with the 'interest driven science' (Grove-White, 1993) of anti-toxics experts. The Derry/DuPont case is interesting in this regard, as well as it marks the first Irish appearance of Dr. Paul Connett. It is also of interest to the GSE's case to consider the manner in which movements mobilised their campaigns around issues, while examining the impact such mobilisation strategies had on particular campaigns

The chapter then goes on to detail the emergence of disputes about landfills in Ireland. Ireland has been traditionally over-reliant on landfill as a waste management solution (Fagan, 2003), and the state's attempts to locate new dumps in areas around the country were met with local resistance. Communities were concerned about the health-risks posed by landfill sites, and groups in Tralee, Ennis, Cork and Galway have challenged the state's landfill policy. In Galway, the challenges against the Carrowbrowne and Ballinasloe landfills would impact on GSE's case, as these landfills reached near capacity and the state responded to the rejection of the CWP by councillors on the grounds of concerns about incineration. However, the potential health and the emergence of environmental groups as risks posed by incineration would be framed as a societal grievance by GSE in 2001 in a similar way to the environmental disputes of the 1970s and 1980s.

In short, the first major political opportunity presented to environmental movements was the state's attempts to locate multinational industries in areas around high unemployment spots. While the Republic of Ireland would go through an economic transformation from the mid-1990s on, the overriding concern of the state throughout was to attract multinationals. As economic stagnation gave way to growth, an emerging consumer society saw the focus switch from industrial to consumer waste.

Baker (1990) has divided the strategic approaches of Irish grassroots groups into 'formal' groups using legal frameworks and 'direct action' groups using pickets and marches. My study indicates that Irish environmental groups have combined both 'formal' and 'direct' approaches to challenge the state. As economic demands shaped the response of the state during these environmental disputes, these movements were unable to make the transition defined by Szasz from 'NIMBY to radical environmentalism' (Szasz, 1994, p. 69). However, the chapter uses Tarrow's understanding of 'four most salient changes' (Tarrow, 1994, p. 85) to explore the manner in which Irish environmental movements have mobilised and exploited opportunities.

3.1.1. Multinational-Led Industrialisation: Pre-'Celtic Tiger' Economic Patterns, 1958–1980

The Republic of Ireland emerged from colonial domination to become a semi-peripheral dependent and agrarian economy on the edge of Europe. Its economy was characterised by high rates of unemployment, and economic stagnation, with little or no indigenous industry, outside of the larger cities such as Dublin and Cork. In 1958 a policy of multinational led development was prioritised by the Lemass government. This policy, devised by the senior civil servant Ken Whitaker, was introduced to help expand the Irish economy, based on the adage 'a rising tide lifts all boats.' In other words, foreign investment through US multinationals would create enough of a wealth-base to enable the Irish economy to develop and modernise. Transnational corporations (TNCs) would provide the investment capital that the Irish business class were unable or unwilling to supply (Allen K, 2000, p. 17).

Although the Lemass/Whitaker economic plan was based on the introduction of direct foreign investment, Ireland lacked an investment class, and never achieved the 'value-system shift' (*ibid.*) necessary to overcome local mistrust of technology, science and multinational industry. Although DeValera's protectionist economic policies were being abandoned, local communities and business interest remained inward looking in the main.

In the aftermath of Ireland's entry into the EU in 1973, foreign investment increased in the following six years by 27% per annum (O'Hearn, 1999, p. 39). These electronics, chemical and pharmaceutical multinationals were primarily US based, with plants being established across Ireland eager to avail of the EU's duty-free tax regime (*ibid.*). The Industrial Development Authority (IDA) was given the remit to attract foreign investment into Ireland. So successful was the IDA that the state became dependent on its ability to continually bring new multinationals to Ireland to create jobs, rather than in promoting indigenous manufacturing.

'As time went on the southern state relied more and more heavily on attracting greater numbers of new firms each year to meet its employment targets' (O'Hearn, 1999, p. 40). The Irish state relinquished many regulatory procedures to facilitate the attraction of inward investment, as well as introducing low corporate tax policies, all of which increased foreign investment, while weakening indigenous business and investment regimes. This loss of control over aspects of economic power has been identified as part of the dependent nature of 'recent rapid growth economies' (O'Hearn, 1999, p. 116) such as Ireland's in the 1990s.

While Ireland's entry into the EU brought some initial benefits, the 1980s brought increased economic stagnation, characterised by rising unemployment and emigration. The Irish state was desperately attempting to attract multinationals into the country at a time when US pollution controls were being tightened, in the aftermath of the Love Canal controversy (Szasz, 1994, p. 5). One result of the Irish state's carte blanche approach towards multinationals was the location in Ireland of industries which became the focus of community campaigns. These campaigns focused on the pollution and health risks posed by some multinationals.

I examine three such campaigns in the 1970s and 1980s. In so doing, the study outlines the shifts in Irish environmental campaigns against toxic industries. These shifts embraced the anti-multinational and NIMBYist concerns of local communities and small business lobbies. The common problem was the toxicity of the products and emission or dumps associated with the relevant multinational industry. This problem of toxicity was framed by local activists as posing health risks for local communities, as well as threatening local farming and fishing industries. In each case, populist responses to the threat of pollution by multinationals represented the beginning of an anti-toxics collective action tradition in Ireland, which GSE could tap into as part of its mobilisation of internal and external resources and opportunities. The following cases show the involvement of educated middle-class professionals, less dependent on multinationals jobs or the State, in creating a backlash against state policy when health-risks are presented and framed in an effective manner.

3.1.2. 1970s: Communities against the Toxic Industries

The state's attempts to site multinationals in locations around Ireland have been contested since the 1970s. Baker (1990, p. 47) has highlighted the nature of Ireland's 'dependent industrial development' that sees the state attempting to locate toxic industrial plants in areas of high unemployment. She points to the peripherality of Ireland's economy which results in highly polluting plants that are unpopular in core states where environmental groups are highly organised. While a consensus on social partnership and neo-liberal economics has been credited with reviving a troubled economy with high unemployment rates (Allen, K. 2000, p. 11), the state's attempts to tackle local unemployment by attempting to lure in foreign TNC's precede the Celtic Tiger. There were two strands of environmental controls that were dealt with by the Minister for the Environment, while the implementation of environmental regulations was the responsibility of the Local Authorities. Compliance with environmental regulations was often negotiable, and 'it was not uncommon for close consultation to take place between the regulator and business interests' (Taylor, 2001, p. 10). This form of consensual clientelism was the norm in the pre-'Celtic Tiger' Ireland, with the focus of local politics fully geared towards the provision of industry and employment. In this context, environmental regulations were not necessarily a priority; indeed the opposite may have been the case:

> There is a problem, in that before an officer from a local authority returns … after taking a sample from a farm or enterprise, some Fine Gael or Fianna Fail councillor or perhaps TD will have made representations by telephone not to proceed with a prosecution (Gilmore, Dáil Debates, 1990, Vol. 394, p. 637 in Taylor, 2001).

Of course, the application of stringent environmental regulations might have jeopardised attempts to attract direct foreign investment into Ireland. With little regard given to environmental regulations, Ireland became a 'pollution haven' (R. Allen, 2004, p. 3). In particular, Ireland became a favoured destination for US multinationals who hoped to avoid the stringent regulations imposed by the US government. While EU environmental directives were adopted as part of domestic legislation, they were not enforced with any degree of stringency, as many multinationals came to believe that:

> All they had to do was fly in, do a deal with local elites, sign a few papers and build a drugs factory (R. Allen, 2004, p. 4).

Environmental issues were not as prominent during the lean economic years of the late 1970s and the 1980s. During this time Ireland was

> near the bottom of the league in terms of concern for the environment, especially when the issue was a trade-off between economic growth and environmental protection. (Whiteman, 1990 from Coyle, 1994, p. 63)

While the Irish Green Party began to make some impact on the national political scene, winning its first seat in 1989 (*ibid.*, p. 65), the state's prioritisation of multinational-led development continued to be subject to challenges from 'locally based indigenous groups' (*ibid.*, p. 63). Local debates about the economic benefits of having a multinational sited in a community as opposed to the potential risks to health and the local environment often led to environmental campaigners being:

> met with hostility within their own community where they have been perceived as anti-development and anti-industry, (*ibid.*, p. 65).

There was already a consensus between the agencies of the state, local authorities, corporations and unions that pre-dated the social partnership ethos that became prevalent in the later 1980s. With the trade unions on board, the only effective opposition to neo-corporatist orthodoxy was found in local community groups, who began to protest about the potentially harmful effects of the pollution emanating from multinational plants. This original consensus, which emerged as the model for the neo-corporatist partnership in the later 1980s and throughout the 1990s, also led to the creation of political opportunities for community groups with concerns about the environmental degradation caused by multinational plants. Environmental movement responses to the state's industrial policy began in the 1970s. Objections to toxic industrial occurred at Pfizers in the early 1970s, Akcan in 1974, Schering Plough in 1974 and Beechams in 1977. Baker has claimed that the focus of these disputes:

> challenges directly the very foundation of Irish industrial policy: development through the attraction of direct foreign investment in Ireland. (Baker, 1990, p. 76)

To highlight the manner in which these environmental groups exploited the shifting political opportunity structure that surrounded the issue of institutional exclusion of environmental actors due to the neo-corporatist orthodoxy, the next section examines some of the historical episodes of environmental dispute which occurred in the 1970s and 1980s in Ireland, as community groups challenged the state and the multinational sector on issues such as the siting of toxic plants, or landfill dumps. This historical survey using Tarrow's (1994) 'four most salient changes' (p. 85) provides an illustration of how the political opportunity structure surrounding environmental challenges to the state has been constituted, and what influence it has had on the mobilisation of such challenges.

3.1.3. Cork: Anti-Multinational Disputes

This section explores campaigns against multinationals that occurred in the 1970s in Cork city and county. Tarrow's 'four most salient changes' (1984, 1988) are applied to these cases, to establish the extent of the access achieved or the alliances that were mobilised during these disputes. The degree of 'ideological development' (Szasz, 1994) undertaken in these cases is also examined. These Cork cases are outlined in the context of a path dependent 'issue history' (*ibid.*) of the Irish environmental movement, which progressed throughout the pre and post boom years, leading up to the current disputes about incineration.

The state's policy in the 1970s of promoting growth through direct foreign investment led to siting of many multinationals in Cork, which had high unemployment rates compounded by the loss of many manufacturing industries in the area. One resource that the state utilised was Cork Harbour, which became the base for an Irish Development Agency industrial zone. This industrial zone would become the location for many chemical and pharmaceutical multinationals (Baker, 1990, p. 58). The IDA's attempts to bring multinationals to the Cork Harbour area coincided with a phase where US multinationals were relocating in the wake of regulations against pollution, following the Love Canal controversy (Szasz, 1994). While US environmental legislation was less stringent than that of the EU,

Ireland's 'open-door' policy of multinational-led development led to it becoming a destination for 'regulation flight' (Yearley, 1996, p. 72), as multinationals availed of the state's prioritisation of industrial led development over regulatory pursuit (Taylor, 2001, p. 26).

Following growing public concern about the safety of the asbestos industry in the United States, increased health regulations forced some US multinationals to seek alternative sites for their plants. The relocation of the Raybestos Manhattan plant to Ovens in Cork is a case in point. This coincided with the IDA's attempts to attract industries to Ireland in the mid and late 1970s, and it led to the opening of the Raybestos Manhattan plant in Ovens, Co. Cork, in late 1975. While the factory began production without opposition, the issue of dumping asbestos waste had become an issue of contention for Ovens' residents by July 1976. Raybestos responded by announcing an improved filtering systems as well as gaining planning permission for a dump (Allen & Jones, 1990, p. 48).

Here, two of Tarow's 'salient changes' in the POS are manifested, as the 'unstable alignments' of Raybestos, the local council, and the State began to become undone as a result of pressure from the resident's group. While Raybestos felt they would be able to open a dump at the plant, local opposition began to drive a wedge between the councillors and Raybestos.

The Ovens residents went on to appeal the dump's planning permission. This allowed the protestors to frame their campaign around the health hazards of asbestos and thus avoid accusations of opposing job-creation in the traditionally hard-hit areas around Cork Harbour (Allen & Jones, 1990). Although there was a degree of consensus in the state and local authorities assessment of the need to attract any form of industry to the Cork Harbour area, the Ovens protestors were also able to exploit the 'unstable alignments' that existed between national and local political entities. While state policy was focused primarily on attracting multinationals such as Raybestos Manhattan under any conditions, local politicians were more vulnerable to the challenge of the Ovens residents, as the health-risks surrounding the operations at the plant became apparent.

Strategically, the residents used public meetings and a petition signed by 600 residents to mobilise their campaign against in the Raybestos plant. However, opposition from by both the trade unions and the Minister for Transport and Power, Peter Barry, soon followed. The residents were accused of 'endeavouring to impede the economic development of the area' (Allen & Jones, 1990, p. 97).

The resident's groups further mobilised their campaign through a rates strike, further pickets, withdrawal of children from local schools and the targeting of waste transport contractors. The protestors also presented a defence of their experts and sources, who had been dismissed by the IDA. A hearing into the dumpsite heard that Raybestos 'proposed to dump 400 lbs. of asbestos in Barnahely per week' (*Cork Examiner*, 30 September 1977). This contestation of science through the expertise provided by challengers was to become a feature of Irish environmental disputes, including that concerning GSE.

After a protracted dispute, 'Raybestos Manhattan had had enough' (Allen & Jones, 1990, p. 119). On 22 October 1980, the company announced its intention to close the plant and cease operation in Ireland, blaming not the protests, but a lack of demand in the US market. The IDA and Irish government were praised for having 'extended all possible facilities' (*ibid.*) to the company. The Raybestos protest then is interesting due to its division between concerned residents' groups and environmentalists on one hand, and the vested interests of trade unions, local and state authorities and the industry lobby on the other. As such, it represents a grassroots challenge to formal political and economic thinking, at a time and at a site of high unemployment. While people were in need of work, they did not wish to suffer health risks or environmental degradation in order to maintain employment. However the Raybestos case:

> showed how far the state was prepared to go to force an industry on a recalcitrant community and how...the state and its various organisations defended and protected the multinational. (Allen & Jones, 1990, p. 120)

3.1.4. The Merrell Dow Dispute

Cork was the site for further disputes of this nature, such as the one based around the Merrell Dow pharmaceutical plant in the Womanagh Valley in 1988. The dispute was based on concerns about the threat to local fisheries and farming in the area around the proposed plant in Killeagh, Co Cork, near Youghal. A challenge to Merrell Dow's 30 million pound chemical plant was undertaken by local farmers, concerned about the affluence from such a massive factory (Peace, 1993, p. 110). This local opposition attempted to gain 'increased access' (Tarrow, 1994, 1998) by forming the Womangh Valley Protection Association (WVPA), in an attempt to have the planning permission for the plant removed, by taking their challenge to a public hearing on the issue in December 1988 (Peace, 1993, p. 110).

The WVPA also sought out the 'influential allies' (Tarrow, 1994, 1998) through their attempts to open up links with 'scientifically qualified' (Peace,

1993 cited in Allen, 2004, p. 49) experts who could lend support to their campaign. An independent scientist, Rory Finnegan, was enlisted to provide this support (*ibid.*). The groups were also able to forge alliances with writers and media figures, who provided publicity for their campaign (Allen, 2004, pp. 50, 51). Alliances were also formed in the locality, with local fisheries and tourism interests (Allen, 2004, pp. 59, 160). The WVPA's approach to resource mobilisation included the public meetings and petition gathering tactics that would later by employed by GSE. However, the WVPA were unable to achieve the 'increased access' or exploit 'divided elites' (Tarrow, 1994, 1998) to the extent GSE would do. Ultimately, it would be the community that would become divided, on the issue of 'jobs versus the environment' (*ibid.*).

After a public hearing where the interest driven science of the WVPA's experts contested the scientific data of Merrell Dow, a series of concessions were offered by the multinational around the areas of environmental monitoring and possible damages (Allen, 2004, p. 69). Merrell also proposed a community consultative committee, to oversee these issues. The planning permission was upheld by An Bord Pleanála, and the WVPA called for a judicial review, as they felt their objections were not dealt with by the hearing. The ruling by Justice Barron in July 1989 stated that the legal basis of the objector's case was open to political interpretation and application (Allen, 2004, pp. 78, 79) and refused to overturn An Bord Pleanála's decision, which was later appealed. During this time, Merrell Dow established its own links with pro-business groups in the Womanagh Valley area (*ibid.*, pp. 82, 83). As the WVPA continued its campaign of public protests and meetings, they attempted to forge alliances with politicians. By September 1989, Merrell Dow announced that the plant would not be built, after representations by the government about the case (*ibid.*, p. 87).

The health risks in this case were based on sulphur emissions, which would affect the water table, farms and local populations. Again, two competing sets of 'official and interest driven' scientific expertise were utilised, and a series of hearings ensued. What distinguished the Raybestos and Merrell Dow cases was the existence of European Economic Community (EEC) environmental legislation in the Merrell case, which called for industries, to be monitored for pollution levels. The existence of EEC directives gave environmentalists and residents' groups a more coherent and legally binding platform to launch legal challenges against multinational polluters. To a considerable extent, the existence of European legislation provided the challengers with the 'influential ally' (Tarrow, 1994, p. 88) necessary as a resource to achieve 'leverage against elites or authorities' (Tarrow, 1994, p. 85). This intervention by the state, in this case, demonstrates a contradiction with the established understanding of the state as having a primarily pro-business stance, to the detriment of environmental concerns. This position would change in the post EPA and neo-corporatist political climate of the years after the Celtic tiger boom.

Irish environmental movement resistance to toxic industries took on certain characteristics at this time. Movements tended to be community based, as multinationals favoured green field sites in areas of high unemployment. By and large, old left institutions, such as the trade unions, and left-wing parties, supported multinationals due to their employment potential, and the constraints they faced due to the fears of wage embargos, which could be enforced by the state. Based on these two cases Irish environmental groups tend to hace a NIMBYist (Not in My Backyard) perspective, as and part of a response to local threats to health and environment. Such movements tend to be suspicious of formal political processes. Likewise the response from the state has been to view such groups with unease and suspicion. The state's prioritisation of competitiveness and growth created political opportunities for environmental movements to exploit, as industrial plants were sited in communities who had concerns about the possible health risks such plants might pose.

Movement participants are often marginalised by local authorities, or by groups in favour of multinational development. They are often dismissed by local political and economic interests as either self-interested locals or outside 'blow-ins', a term often applied to Irish residents originally from the United Kingdom, Europe or North America who may bring wider environmental understandings to local issues. The official response to community-based resistance was to centralise the planning process, and exclude locals from local development initiatives as much as possible. It could also be said that Ireland's colonial past has left a residue of suspicion and hostility amongst local communities in regard to attitudes towards the mechanisms of the state (Tovey, 1992) and in the state institutions themselves, for example, Office of Public Works (OPW).

3.1.5. 1980s: The Rural Response: The Hanrahan Challenge against Merck Sharp and Dohme

When the Merck Sharp and Dohme chemical plant opened in South County Tipperary in 1976, it was seen as a victory for the IDA, who had joined with South Tipperary County Councillors in defeating a small group of environmentalists who had objected to the plant. During the debate about the potential benefits of the plant, 1,000 jobs were promised. However, only 260 jobs were delivered. With regard to the costs, it is necessary to point out that EEC environmental policy had not yet been clearly established in relation to industrial pollution. This meant that the Merck Plant had few limitations placed on its pollution capacity.

Within two and a half years of the Merck plant commencing operation, the family had reason to suspect that health problems, animal and human, were being caused by emissions from the factory (Allen & Jones, 1990, p. 35). The Hanrahans began to frame their challenge to Merck Sharp and Dohme with the assistance of the 'interest-driven science' provided by the Canadian toxicologist Dr. James Neufield, who contested the data on dioxins provided by the Department of Agriculture.

Eventually, the Hanrahans took a High Court action against Merck Sharp and Dohme. A community group was formed to object to the granting of permission for an extension to the plant. Picketing of council meetings continued through 1984, as the symptoms of poisoning continued to be a concern for local farmers. The protests continued despite local economic concerns:

> Merck contributes an estimated £50,000 a year to health, education, environmental, civic and social projects all over the country, with a significant amount going into the local community. (Allen & Jones, 1990, p. 39)

Despite this, local residents combined to form the Suir Valley Residents Protection Association in 1984. Despite the objections of the group, planning permission was given for a further extension to the Merck plant. A further development saw John Hanrahan take a High Court lawsuit against Merck. The trial detailed the suffering endured by the Hanrahans, which included the loss of 225 animals that they claimed had died as a result of the emissions from Merck between 1980 and 1984. Merck countered this claim with an accusation of farm mismanagement, a claim that was supported by officials from the Department of Agriculture. Dr. Neufield provided evidence of a similar case to the Hanrahan's in Canada. His presence at the trail provided the 'influential allies' necessary to present an interest driven scientific challenge against Merck. The case also revealed that the health of the Hanrahan family members had suffered from symptoms similar to that of toxic poisoning (Allen & Jones, 1990, p. 40), due to the support provided by the local veterinarian. By mobilising the resource of expertise, the Hanrahan case was able to challenge the orthodox arguments that were presented by the state, in a framing process that focused on the toxicity and health risks facing the family. However, these allies were not to prove 'influential' enough to 'increase access' for the Hanrahans.

At the end of the case, in July 1985, the High Court ruled against the Hanrahans. Legal costs were set at £1 million and credit was removed and farm equipment auctioned, to pay costs and fund an appeal. The appeal was heard in 1987, and this time the ruling was in favour of the Hanrahans. The undertaking of the Merck incinerator was highlighted as a probable cause of dioxin emission. According to the Justice Henchy's summation:

> The most credible explanation offered for the ailments and abnormalities in the earth was the toxic emission from the factory. (Allen & Jones, 1990, p. 44)

The Hanrahan case demonstrates the ability of a local family, with community support, and interest driven 'influential allies' to take on a multinational toxics industry, by exploiting the resources and political opportunity available to them. This was achieved by the success of these 'influential allies' as the court action taken by the Hanrahans allowed scientific experts to challenge the 'official science' (Grove-White, 1993) of the Department of Agriculture. The use of scientific expertise allowed the challengers to frame the issue in a manner that maximised the sympathies of the public for a local farming family, as the health risks posed by Merck's plant were highlighted. The family of issues round toxic plants would play a major part in the outcome of a subsequent grassroots campaign against the toxics industry, that of the Raybestos Manhattan Plant, in Co. Cork.

By applying Tarrow's (1984, p. 86) 'four most salient changes' to this rural dispute we can identify key points in the 'dimensions of opportunity' that surrounded the Hanrahan case. By linking the Hanrahan's own grievance with the concerns of the farmers and small indigenous business interests of the Suir Valley, 'increased access' (*ibid.*) was achieved. 'Influential allies' (*ibid.*) were found through the involvement of 'interest-driven' scientists (Grove-White, 1993, p. 21) from Canada.

3.1.6. The Cross-Border Response

In this section (see also Leonard, Doran & Fagan, 2010; Rootes & Leonard, 2010), Tarrow's 'four salient changes' will be applied to a cross-border case. While the economic dependency of the Republic of Ireland had shifted from a reliance on Britain to a policy of attracting US multinationals (O'Hearn, 1999, pp. 51–54), Northern Ireland had made no such transition, remaining 'subservient' (*ibid.*) to the wider UK economic circle. Northern Ireland's economy went into decline in the 1970s and 1980s, as 'a lukewarm distant government' (*ibid.*) remained slow to invest in the strife-torn province. More importantly, Northern Ireland was unable to avail of the type of focused drive for direct foreign investment that had been undertaken by the Irish state (*ibid.*). As a result, two very different economics had emerged, North and South, by the 1990s. This dichotomy is illustrated by the fact that:

> 629 non-British transnational companies employed 73,800 in the South compared to just 86 such subsidiaries employing 17,826 people in the North. (Hamilton, 1992, p. 86 cited in O'Hearn, 1999, p. 54)

One multinational that attracted support from both the Irish state and the Northern Ireland Office (NIO) was the DuPont plant in Derry. Both the Republic's Minister for Environment Padraig Flynn, and the NIO's Department of Environment were involved in negotiations with DuPont to open an all-Ireland toxic dump (Allen, 1992, pp. 4–23). While DuPont had received planning permission for a 'chemical waste burner' from the North's Department of Environment as far back as 1981 (*ibid.*), it was Minister Flynn's negotiation with DuPont to locate an all-Ireland toxic waste incinerator at the Derry plant which provided the political opportunity, as well as the 'unstable alignment' of the two authorities on the island, for a community-based challenge to the incinerator as well as the industrial policies of both jurisdictions.

As the concerned citizens of Derry mobilised, they were able to draw on the existing experience and expertise of the Cork Environmental Alliance that had been active against Raybestos Manhattan (Allen, 1992, p. 5) The proposed incinerator was to take waste from both sides of the border, leaving the Derry campaigners two sets of government officials with which to engage. Minister Flynn's proposal to send toxic waste from the Republic to the North has been called 'a Northern Irish solution to a southern Irish problem' (Allen, 1992, p. 5). The Northern Ireland Department of the Environment also became involved in the project. Opposition to the incinerator also took on a cross-border manifestation, as the campaign drew on two groups; the Inishowen Environmental Group from nearby Donegal, and the Derry-based Greencastle and Moville Environmental Group.

Experts were drafted in by both sides to support the pro or anti-incinerator argument. Questions were raised about the need for a cross-border incinerator, what type of toxics would be burned there and what were the subsequent health-risks, and whether DuPont had a history of environmental pollution There was also the question of how would they respond to a health problem at their Derry plant (Allen, 1992). An anti-toxics rally in Derry on 16 March, 1991, led to further clarification on the basis for concern about DuPont. Protestors were anxious to pinpoint what harm the process of incineration could do to local residents. Again, the community groups turned to 'influential allies' with scientific expertise. Interest based scientific experts were consulted, including Dr. Paul Connett, spokesperson of the US anti-toxics movement. In interview, Connett, who was to become prominent in GSE's case stated that

> My first incinerator battle (in Derry) was the DuPont incinerator. Then the next year we were back in Derry, in Donegal, fighting trash incinerators, people forget this, we keep going through the same arguments again and again .... (Dr. Paul Connett interview, 21 January 2002)

The expertise provided by Connett and others provided the basis for the anti-toxics groups' opposition to DuPont, and he helped the Derry protestors to frame the dispute around the potential for health risks posed by DuPont, exploiting the fears of the public. 'Influential allies' could also be found in local politicians, such as by John Hume, the Social Democratic Labour Party (SDLP) leader, who was particularly concerned about the importation of waste into Derry.

With the change of government in Ireland in 1992 and the rise of the peace process as a focus for northern Irish politics, financial aid was not forthcoming. The Derry anti-toxics groups framing of the health issue had proved to be successful, as their presentation of 'interest driven science' (Grove-White, 1993, p. 21) led to the postponing the plans for a national incinerator, and in establishing the anti-toxics movement in Ireland. The 'mix of open and closed factors' (Eisinger, 1973, p. 15) such as the history of oppositional politics in Derry and the 'unstable alignment' of the NIO and the Dublin government led to 'increasing access' (Tarrow, 1994, p. 36) for the Derry community challengers to exploit.

The DuPont dispute can then be analysed through the application of Tarrow's 'four salient changes' (1984, p. 86). 'Increased access' (*ibid.*) was created by the Derry campaign's ability to create links with 'influential allies' such as Dr. Paul Connett, who later became a central figure in the GSE case. By introducing Connett and his 'interest-driven' science to the campaign, the Derry protesters were able to challenge those in favour of the DuPont Plant, including the Irish government and the Northern Ireland office. Another 'influential ally' was the SDLP leader John Hume, who provided political support for the campaigners.

3.1.7. Conclusion

The Merck, Sharp and Dohme, Raybestos Manhattan and DuPont protests represent the emergence of a series of challenges to the state's policy of multinational led development. Despite high rates of unemployment and emigration, some sections of the community were unwilling to accept toxic industries in their areas. The biases facing such groups were immense, while the mobilisation of resources such as financial support, political alliances and professional expertise were limited. As environmental campaigns continued in the 1980s and 1990s, the dynamic of the POS evolved, and challengers became more adept at exploiting opportunities and framing issues. By 1992, the EPA had been established, although it has been criticised for being subordinate to the IDA (Taylor, 2001).

The state would soon be faced with another environmental issue, that of the growing waste crisis in the 1990s. While some of these environmental challenges led, in some way, to the closure of the toxic plant in question, others have been less successful, as the 'strength of the state' (Tarrow, 1994, p. 89) reduces the capacity for challengers to exploit political opportunities to the fullest. According to Tarrow 'strong states also have the capacity to implement the policies they choose to support' (Tarrow, 1994, p. 89) limiting the effectiveness of environmental challengers. As the Irish State came to centralise power through its neo-corporatist partnership arrangements from 1987 onward, it would become apparent that movements would need to expand their strategic approaches.

3.2. THE 1990s ECONOMIC BOOM: THE 'CELTIC TIGER'

The 'Celtic Tiger' economic boom was a period of accelerated economic growth, followed by a construction led property boom, which lasted between 1996 and 2004 in Ireland. It was ended by the global downturn and domestic banking related fiscal crisis that occurred in 2008. While the GSE dispute had its basis in a consumerist driven waste-management crisis that had emerged in the years subsequent to the mid 1990s economic boom (Tovey & Share, 2000, pp. 480–482), the preceding decades had been marked by a degree of economic 'pessimism' (Taylor, 2002, p. 1) due to the ongoing stagnation in growth and high rates of emigration and unemployment.

To secure a successful economic performance in Ireland has been the overriding policy agenda of the state since the late 1950s. Throughout the next four decades, successive governments would attempt to kick-start the Irish economy by attempting to entice TNCs to locate in Ireland. One of the features of the state's sales-pitch to TNC's was a low corporate tax regime combined with an ambivalence at local authority level to pollution at the source from newly located industrial plants. Local authorities were also hindered by the lack of funding for regulatory surveillance, as well as lacking in the necessary expertise for the monitoring of pollution (Taylor & Murphy, 2002, pp. 81–82). This situation was further complicated as:

> local authorities were, on the one hand, charged with monitoring and regulating the environment and yet, on the other hand, were in competition with one another to court the investment of multinational companies. (*ibid.*)

The phase of economic growth that was experienced in Ireland in the past decade came after three decades where government policy was primarily focused on industrial development and job creation. By opening the Irish economy to globalised market forces, a period of intense economic growth was achieved, often referred to as the 'Celtic Tiger' (O'Hearn, 1999). One result of this growth in industrialisation, urbanisation and consumerism was seen in the emergence of a waste management crisis across Ireland. The Irish government also had to operate within the constraints of the wider political demands of the EU. This led to the directives and legislation on pollution controls, which changed the nature of environmental disputes in Ireland, from a focus on health risks to one that was concerned with the waste management crisis that resulted from rising consumption rates.

As rates of affluence and consumption increased (Allen, 2000, p. 68), Ireland's waste-management infrastructure (which at the time consisted in the main of municipal landfill sites) was unable to cope with the increased rates of waste being produced. At the same time, the demands of European Commission directives on waste management called for new approaches to waste management in Ireland. Article 5 of the directive says that member states should aim for 'self sufficiency in waste disposal'. This placed landfill at the bottom of the scale, and greater emphasis was placed on reducing, reusing and recycling materials. Incineration was also an option according to the waste hierarchy, but Ireland has not introduced mass-burn incinerators up to this point. In 1992, the government established the EPA to regulate waste disposal. Its remit has been criticised for not focusing on environmental degradation, and for its subordinate role to the IDA and the project of industrial development. The emergence of the EPA and Irish legislative framework for waste management will be explored in detail in Chapter 4.

According to Taylor:

> while the Irish State had recognised the need to revamp environmental policy, its principal overarching objective was to ensure that further more stringent regulation would not be detrimental to the economic performance of the Emerald Tiger. Put simply, environmental policy debate in Ireland is concerned no longer with the extent of ecological degradation … but with … the extent to which pollution could be justified. (Taylor, 1998, p. 5)

As the demands of EU directives and the state's own legislative directive dictated that landfilling alone was no longer acceptable waste management practice, the state was faced with a new problem, as communities began to protest about the siting of dumps in their areas.

3.2.1. Framing the Landfill Problem

This section examines the manner in which local communities exploited the 'unstable alignments' (Tarrow, 1984, p. 87) that existed between national and local government in relation to waste policy. One critical aspect of the growing waste crisis was linked to the fact that Ireland disposed of up to 93% of its municipal waste through landfilling. Such is Ireland's over-reliance on landfill disposal that the country in ranked second only to Russia in European landfill dependency. Landfilling has fallen to the bottom of the EU's waste management hierarchy, below first preferences such as reduction, reuse and recycling. Concepts such as the 'polluter pays principle' have become central to the EU's environmental policy (O'Sullivan, 2000, p. 21).

Landfills are also facing increased legal challenges from local community groups who fear an appearance of a regional 'super-dump' in their area. Residents' groups in Tralee, Ballymahon (Longford) Doora/Ennis, Ballyguyroe, Co. Cork, as well as in Carrowbrowne and Ballinasloe, in Galway have taken High Court proceedings, exploiting the 'unstable alignments' that existed between the state and local authorities in relation to waste policy. These court actions have led to orders against the relevant local authorities forcing landfills in these areas to close, or to close after a period of time (O'Sullivan, 2000, p. 21).

To contest the state's landfill policy, these community groups have framed their challenges around nuisance factors, health risks, agricultural impacts and legal breaches. The first of these cases occurred in Tralee in 1992. Two Tralee families, the Baileys and the Galvins, took an action against Tralee

Urban District Council that criticised the operation of the Tralee landfill. Later that year, a dispute over the Ballymahon landfill in County Longford led to a more protracted case. Local residents picketed the site, with the result that Longford County Council obtained an injunction against the residents. In turn, the residents obtained an injunction requiring the closure of the Ballymahon landfill site. After appeals to the High Court by both parties, a negotiated settlement led to the closure of the landfill. Residents' legal costs were also paid (*ibid.*). The main framing issues for the anti-landfill groups included

- the health risks: posed to communities through water table and seepage pollution, increased rodent and vermin infestation and unbearable odours from dumps.
- the agricultural and tourism/visual impacts: due to land erosion, toxicity levels from dump residue and the impact of large-scale landfills on the landscape.
- the legal framework: community groups were able to utilise EU legislation to support their position, as landfill was the least-favoured option of the EU's waste hierarchy.

In 1997, the Kildorrey Anti-Dump Group took a High Court action against Cork County Council, claiming that the council had failed to comply with the legal operating procedures of the Minister for the Environment. A lengthy case led to a negotiation on landfill operations, a date for closure, and an abandonment of plans for expansion of the site. Costs were also awarded to the group. The following year, residents from the area around Doora, the Ennis dumpsite in County Clare, took a High Court action against Clare County Council. The residents framed their concerns about landfill claiming the dump was causing:

> significant nuisance smells, visual intrusion, noise, fires, smoke, ash, wind-blown litter, water-borne litter, rats, flies, birds, flooding of adjacent lands and damage to agriculture. (O'Sullivan, 2000, p. 21)

These problems had led to a decrease in the locals' quality of life and to property prices, which can be seen as giving rise to NIMBYist concerns. The council was accused of causing these problems, due to poor management of the Doora site. A 23-day hearing at the High Court led to negotiations on the operation and closure of the site, with no expansion allowed for the future. Costs for the group were settled, as was compensation. Having framed the issues as one of environmental degradation, rather than focusing only on property prices, the residents were able to succeed in presenting themselves as pro-environment rather than NIMBYist.

In the same month, three residents from the Carrowbrowne area of Galway took a case against Galway Corporation, regarding their operation of the Carrowbrowne dump. The residents again framed the issue in a particular way, citing legal breaches as the basis for their challenge. Galway Corporation had obtained permission to operate the dump for two years from April 1997, with a number of conditions attached. The residents claimed that these conditions were being ignored. Judgement was for the residents, with Justice Kelly claiming that the Corporation was:

> involved in the commission of a criminal offence ... in serious breach of the planning laws; while the Corporation's approach to the case 'was quite alarming' and 'leaves a lot to be desired'. (O'Sullivan, 2000, p. 23)

Landfilling was prohibited by order 'as of today' (*ibid.*), according to Justice Kelly. An extension was granted to facilitate Christmas collections, after which Galway city's waste was to be taken to the Poolboy dump at Ballinsloe. This in turn led to locals to picket the Poolboy dump, with injunctions being taken out by the Urban District Council to get residents to lift the picket. The Poolboy residents subsequently obtained an injunction against Ballinasloe UDC, repeating the procedure of previous anti-dump campaigns. A settlement was negotiated through the High Court, which will close the Poolboy, Ballinasloe dump in 2005. Poolboy is to be run in accordance to EPA practices in the meantime (*ibid.*). While the landfill issue has been met by many similar challenges, it is interesting that these community campaigns have framed the landfill issue in a manner which has created a moral and 'orthodox consensus' (Grove-White, 1993, p. 19) which runs parallel with the demands of EU directives on landfill. It is in this context that the state's regional waste policy can be viewed.

The European waste hierarchy has emphasised waste reduction at the source. Reduction, reuse and recycling are the preferred options while incineration and new forms of landfilling remain contentious. Irish governments are feeling pressure from above to comply with EU environmental directives, in addition to pressure from below from community groups who oppose landfill any new site designated for incineration. The detail of the 1996 Waste Management Act also emphasises the responsibility of local authorities in relation to the planning for, and implementation of, new waste management practices as presented under the Act. Despite the acceptance by government of the 'best practices' approach, local authorities continue to approach new waste management practices with a degree of reluctance.

> Very few of the waste management plans produced by local authorities since 1997 give priority to these waste management strategies. Instead, most of these plans discuss all options except landfilling or incineration … and there has been little meaningful public involvement or participation in their preparation. (O'Sullivan, 2000, p. 23)

In contrast, the EPA has received up to 7,500 submissions in relation to waste license applications, reflecting the high degree of interest and concern on the part of a public that has been excluded from the waste consultation process. While the policy statement and regional waste plans contain 'best practices' such as reuse, reduction and recycling, a stark choice between a waste crisis brought on by the closure of landfills and incineration has been presented by the government in recent advertising campaigns. As the 'partial access' (Tarrow, 1994, p. 86) created by the level of public submissions as a vehicle for challengers increases, there has been a groundswell of opposition to incineration across Ireland.

In this section, I have examined the manner in which communities have challenged the state's attempts to locate landfills in their areas. In terms of Tarrow's 'four salient changes (1984, p. 86) we can see that these community groups 'increased access' (*ibid.*) by challenging the 'unstable alignments' (*ibid.*) that existed between the state and local authorities, as the legal system was utilised to prevent landfills being opened, or having their licenses extended.

3.2.2. 2000–2004 Waste Management Goes Regional: The Cross-Border Strategy and Campaigns in Cork and Meath

One of the discourses to emerge as a result of the Peace Process in Northern Ireland was the potential for joint policy initiatives on many of the issues that affected life on both sides of the border. There were many reasons why the 1990s presented new opportunities for increased cooperation on waste management in an all-Ireland context. One significant factor was the increased political cooperation between the British and Irish governments in the wake of the Downing Street Declaration and the Peace Process from 1992 onwards. Another factor that assisted regular cooperation was the establishment of the EPA, also in 1992. A third contributory factor that brought increased cooperation was the establishment of a regulatory framework on waste management by the EU, which both jurisdictions had to comply with. A cross-border study undertaken by the National Institute for Regional and Spatial Analysis in 2001 places the focus for delivering waste management policy on the EU and its directives on:

> dangerous substances, waste oils, grounds water, urban waste water, licensing regulations, the disposal of toxic waste, sewage sludge in agriculture, emissions from incineration plants, the disposal of animal waste, and batteries containing dangerous fluids. (Fagan, 2003, p. 72)

The 2001 study on waste management strategy identifies a commonality between approaches to dealing with waste output north and south

> because of a perceived threat to industry and investment, to the 'Celtic Tiger' in the south and the 'peace dividend' in the north. (Fagan, O'Hearn, McCann, & Murray, 2001, p. 44)

There is an additional difficulty with implementing localised waste policies, as local resistance to plans for landfills, incinerators and reprocessing plants are opposed by communities. The study identifies the need to locate the waste management issue in a sustainability context, 'within the broader framework of all-island social and economic structures' (Fagan et al., 2001, p. 40). This concept incorporates an understanding of convergent policy 'path dependency' (*ibid.*) in relation to implementing an all-Ireland waste hierarchy, which can focus on 'material recovery' (*ibid.*) as well as waste reduction. In so doing, such cross-border policies must be established in tandem, and implemented by the relevant authorities across the island. 'Quick-fix' (*ibid.*) solutions that compromise the potential for cross-border agendas on waste should be avoided, as they could result in a deviation from the drive towards sustainability.

The study finds that two divergent and separate waste policy agendas on the island are unsustainable. It also identifies concerns about incineration being a short-term solution that 'relieves' policy-makers from 'the responsibility for seeking more sustainable approaches to waste management, particularly waste reduction' (*ibid.*). Incineration could 'block other options' and was 'the most expensive option for waste management' (*ibid.*).

The study also finds that a democratic deficit arises from the manner in which waste policy is implemented. While public participation is encouraged in waste legislation, the public are 'excluded at key stages' (Fagan et al., 2001, p. 41) of the policy-making and implementation process. Local authorities appear to be unwilling or unable to incorporate communities into this process and haven't been able to properly assess the depth of local feeling in relation to waste policy projects, whether these are the result of 'negative (NIMBY)' or 'positive (environmentalist)' perspectives (*ibid.*). The authorities have attempted to deflect public resistance to waste policies by 'individualising' the problem, by presenting it as a domestic consumer issue, or by the 'marketisation' or 'economisation' of waste management issues (*ibid.*). These two approaches focus on altering people's behaviour in the domestic or market sphere, urging the public to 'decrease our waste' while making 'sustainable practices profitable' (*ibid.*) for corporations, and shifting the nature of the debate about waste management into one 'about profitability rather than one of responsibility' (*ibid.*).

The study finds that waste management policy should be achieved through 'economies of scale', with an 'all-island dimension' (*ibid.*) to its implementation. This approach would involve the development of markets and plants to assist recycling and materials recovery projects on an all-island basis, with cross-border co-operation becoming the most economically efficient way of proceeding. There is an importance for policy makers to recognise

> the interconnectedness of waste management and other spatial policies for population growth, industrial placement, physical infrastructure and other socio-economic factors. (Fagan et al., 2001, p. 44)

To maximise the potential of such spatial policies, the barriers that exist between both jurisdictions in relation to legal and bureaucratic systems need to be reviewed (*ibid.*).

The study provides some conclusions and recommendations for future waste management strategies. These include the adoption of a 'user hierarchy' to go with the waste hierarchy, to overcome top-down approaches to the waste hierarchy through an emphasis on incineration rather than waste reduction. The user hierarchy would allow for a fairer assessment of individual consumers in relation to the waste produced by the industrial and agricultural sector (Fagan et al., 2001, p. 46), in the context of distinguishing between 'short-term needs and long-term strategy' (*ibid.*). Markets will not be able to produce sustainable strategies, and 'international best practice' can be learned from, while 'seeking where possible local solutions' (*ibid.*). In this way, waste can be seen as a 'global' (*ibid.*) problem, with global and local resonances. EU directives provide the impetus for waste policy and through the acceptance of EU 'subsidiarity' practices, where policy is initially implemented at local level, a maximum recognition and acceptance of policies and best practices can be achieved (*ibid.*).

Acceptance for these policies can also be better achieved through partnerships between local communities, businesses and the authorities.

Acceptance can also be maximised, with greater efficiency, through an 'all-Ireland' approach to waste management, with significant levels of cross-border co-operation. This incorporates both an economies of scale as well as politically inclusive approach that is in accordance with the principles of the Good Friday Agreement (*ibid.*). Fagan and colleagues recognise the 'social dimension' of waste management, which goes beyond the domestic sphere, and restricts communities through accusations of NIMBYism is judged to be unfair, and also neglects wider aspects of the problem, such as waste from the industrial and agricultural sector (*ibid.*). Good governance on an all-Ireland level can be achieved through the utilisation of community networks with widespread research on possible options and to provide further understandings of possible contexts that can promote the best frameworks necessary for successful sustainable waste management policy implementation across the island.

The study presents an understanding of the potential change in context for future waste strategies, if current inefficient practices in both jurisdictions can be overcome. Other negative factors, such as some northern concerns about all-island co-operation, or the absence of inclusive measures for local communities in relation to consultation on waste issues, can be overcome if best practices and subsidiarity are properly implemented. Through the one-island approach, local unstable alliances can be overcome, as the EU provides the 'influential ally' necessary for the implementation of sustainable waste management strategies. More significantly for the GSE case, the study is critical of the use of incineration as a short-term or quick-fix solution to Ireland's waste problems.

3.2.3. Cork Anti-Incinerator Dispute: Cork Harbour for a Safe Environment

Influencing the findings of the 2001 cross-border study on waste management were, no doubt, the disputes the late 1990s in Meath and Cork, where sites for incinerators had been announced. These sites, at Duleek, Co. Meath and Cork Harbour, were part of a series of municipal incinerators announced by the state as part of their seven regional waste plans in Dublin, the South-East, Galway, Limerick, the Midlands and the North-East (Fagan, 2003, p. 67). Communities in these areas campaigned against the proposed incinerators, using collective action frames to create understandings about the potential health risks posed by incineration. This section explores the manner in which campaigners in Meath and Cork engaged in the framing process, as well as examining the mobilisation of resources to achieve access to political opportunity structures surrounding the waste issue. The Meath and Cork campaigns had similarities with GSE's case, as political opportunities were provided by the state's approach to including incineration in their regional plans, including changing the word 'incineration' to the less controversial 'thermal treatment':

> One can see that incineration is the contested terrain in this case, as not one government policy or regional plan mentions the word 'incineration.' The word used repeatedly and pointedly is 'thermal treatment plant.' As in all conflicts, the discourse itself marks the terrain and the use of the work 'incineration' as opposed to 'thermal treatment plants' marks the political division. (Fagan, 2003, p. 78)

In other words, the state's own attempt to frame the waste issue is a less than 'up front' (*ibid.*) manner provided the first political opportunity for community campaigners to challenge the state's regional waste plans.

In December 2001, the Belgian incineration company, involved in all the regional plants, Indaver, lodged a planning application for a 100,000 tonnes per annum incinerator for hazardous and non-hazardous industrial waste, as well as an incinerator of equal size for municipal waste at a site at Cork Harbour. This incinerator was opposed by activists in the Cork Harbour/Ringaskiddy area, which had been the site of previous disputes, such as Raybestos Manhattan. This provided their campaign with 'increased access' (Tarrow, 1984, p. 86) due to the existence of many activists who had taken part in previous campaigns. As such, the existence of resources of campaign and scientific expertise could be more easily mobilised by the campaign.

The proposed incinerator was opposed by local residents, who formed an alliance from the many existing environmental groups in the Cork area, under the name of the Cork Harbour Alliance for a Safe Environment (CHASE). The campaign gained momentum, and CHASE were able to claim to 'represent the views of 24,000 people' (*Irish Examiner*, 19 September 2003). CHASE followed similar strategies to Galway for a Safe Environment, by establishing a network of environmental groups, mounting a campaign of public protests and meetings, and even drawing on the support of Dr. Paul Connett. However, CHASE had one major difference in their strategic approach; it focused on a direct challenge to Indaver, rather than aiming to influence the political process. Indaver and the state were keen to site their incinerator at Ringaskiddy, Cork Harbour, as '60% of all hazardous waste in Ireland comes from Little Ireland and Ringaskiddy' (*ibid.*). Links were established between CHASE, GSE and the No Incinerator Alliance in Meath, as resources and expertise were exchanged between the groups.

CHASE had mobilised the concerned residents of the Lower Harbour and Ringaskiddy areas of Cork. Its campaign consisted of public protests and rallies, and the groups framed their campaign around the health risks of dioxins posed by incineration. CHASE had a high number of local women involved in their campaigns, which was similar to previous environmental campaigns in Cork, such as the Raybestos Manhattan campaign (Allen & Jones, 1990). The CHASE campaign saw the appearance of many of the women and children in white boiler suits and oxygen masks, holding 'No to Incineration' placards, and presenting a striking visual profile (*Irish Times*, 17 January 2004). CHASE were able to frame the 'moral discourse' (Grove-White, 1993, p. 20) surrounding the issue, depicting the children in a way which highlighted their potential exposure to the health risks of dioxins.

The dispute was taken to a series of oral hearings, held at the Neptune Stadium in Cork. In this way, the campaign led by CHASE manifested itself in a manner that had become the norm for environmental disputes in Ireland. The role of the oral hearing in Irish environmental disputes has been summarised as follows:

> The EPA Act has implemented a structure in which oral hearings perform an advisory function, a move which sits comfortably within the Irish administrative tradition which prefers an adversarial format as a means of gaining evidence. In such circumstances, the inspector's role is not to produce a decision, rather it is to collate information and report back to the EPA. It is the inspector's task to define what is useful and what is not. In the final report he or she may make recommendation, but the decision ultimately lies in the hands of the EPA (or the Minister). (Taylor, 2001, p. 136).

Owing to the inherent limitations faced by movements which undertake legal challenges which may only provide 'partial access' (Tarrow, 1994, p. 86) and the high costs stemming from the need for legal expertise, the legal process has been favoured by the EPA and the government as a vehicle for settling environmental disputes. It is also the reason such a strategy was avoided by Galway for a Safe Environment. Such strategic decisions provided GSE with their level of success, in contrast with CHASE. After a campaign lasting over two years, An Bord Pleanála (the planning commission) refused permission for the Ringaskiddy waste scheme on the following grounds:

- The Environmental Impact Statement was inadequate.
- A hazardous waste incinerator would be premature, prior to progress on waste prevention targets in the National Waste Management Plan.

- It would be premature with no hazardous landfill site.
- An incinerator for non-hazardous industrial waste was contrary to the Cork Waste Management Plan 1999.
- It would contravene materially the zoning of the site primarily for industry and enterprise.
- It would contravene the county council's objective for large stand-alone industry.
- It would contravene the county plan's objective to safeguard lands at ports and harbours against inappropriate users.
- Because of its bulk, scale, height, design and location, the proposed development would be visually obtrusive.
- The scale, nature and purpose of the development would be fundamentally unsuitable to the site.
- It would be close to high density housing in Ringaskiddy.
- It would generate an excessive movement of traffic.
- There is already serious congestion.
- It would be premature given deficiencies in the road network.
- An Bord Pleanála should not be satisfied that the development would not pose significant risks to public safety (*The Irish Times*, 17 January 2004).

The senior planning inspector who had dealt with the case and presided over the public hearing was Phillip Jones. His report ran to over 300 pages and rejected Indaver's plan for an incinerator on the fourteen counts listed above. CHASE claimed a complete vindication for their position on the issue, as many of their objections formed part of the inspector's rejection of the incineration scheme.

However, both the inspector and CHASE were soon to find that An Bord Pleanála would overrule the refusal and grant planning permission for Indaver's proposed plant, citing the prioritisation of government waste policy over all other considerations. It was reported that the refusal for the Ringaskiddy plant had angered the Minister for Environment, who felt frustrated due to the state's need to have its waste policy implemented in the face of community resistance (*Irish Times*, 17 January 2004). The state's response to the CHASE campaign could be seen as 'exclusive' (Scharpf, 1984, p. 60). Despite the facilitation of the challenges through the public hearing, the state eventually overrode its findings, due to the demand for an incinerator from local multinationals. This understanding of an 'exclusive' response to environmental campaigners has been established on three levels, 'repressive', 'confrontative', and 'polarising' (by Sharpf, cited in Kriesi, 1995, pp. 173, 174).

An Bord Pleanála also had to consider the state's waste management framework, in which a preference is given to incineration over landfill. The board also considered the geographical location of a number of large-scale chemical and pharmaceutical industries operating in the Cork Harbour Area, and stated that Indaver's plant was 'an appropriate location for a necessary public utility' (*Irish Times*, 17 January 2004). It also stated that such a plant would 'not injure the amenities of the area or be prejudicial to the future for port-related development' (*ibid.*).

Despite the justification of the inspector's emphasis on reducing waste as having priority over disposal facilities in line with the EU's waste management hierarchy, An Bord Pleanála agreed to the plant for the following reasons:

- It is in line with the objectives of the National Hazardous Waste Management Plan.
- The proposal for incinerating non-hazardous waste is in accordance with national policy (the plant will deal with both hazardous and non-hazardous waste).
- It is in accordance with the Cork County Development Plan.
- The inspector's findings on visual impact are rejected.
- In considering the effects on residents the board had particular regard to the zoning of the site and the separation from the nearest houses.
- The board had particular regard to the proximity of the site to the N28 and the proposals for upgrading this route.
- On the Environmental Impact Survey (EIS) the board did not consider that significant data was withheld (*Irish Times*, 17 January 2004).

Opponents of the incinerator were stunned at the decision, claiming it was 'a sad day for Cork' (*ibid.*). CHASE spokesperson Linda Fitzpatrick said of the decision:

> We are bitterly disappointed and totally shocked at the ruling. We put two years hard work into this. We put on excellent case together and went into immense technical detail. This decision shouldn't have gone ahead as far as we're concerned. (*ibid.*)

CHASE criticised An Bord Pleanála due to its failure to fully assess the health risks involved and questioned the value of having a planning process if the finding of such procedures are ignored. The decision was also criticised in relation to the lack of transparency with the government's influence on the outcome. CHASE also feared for the future of the Cork Harbour Area:

> The fact of the matter is that Ireland has 15-20,000 tonnes of hazardous waste per annum for incineration. A commercial company has now been approved to build a 100,000 tonne facility. It is unnecessary and wrong. (*ibid.*)

At that point, CHASE began examining the prospect of fielding candidates in the forthcoming local elections scheduled for the summer of 2004. This may have provided the group with a better chance to exploit the 'unstable alliances' that could be said to have existed between the state and local political figures. Opposition politicians criticised An Bord Pleanála for giving into government pressure on the issue. Cork TD and Fine Gael's environment spokesperson, Mr. Bernard Allen, demanded that the board make a full statement justifying its decision and explaining why it had overruled its planning inspector's decision (*ibid.*). Mr. Allen also called on the Environment Minister to explain the decision to Cork County Council. Senator Brendan Ryan of the Labour Party claimed

> We now have a policy that will create an incentive to generate such waste and will create enormous fear and anger in the whole region in the vicinity of Ringaskiddy. As a chemical engineer and a resident of the area, I regard this decision as wrong, both technically, and in terms of good environmental policy. (*Irish Times*, 17 January 2004)

The then Green party TD for Cork South Central, Dan Boyle, said the decision was a 'political' one, and accused the board of 'caving in' to government pressure. He questioned the fairness of the planning process. He said that the campaign against the plant would continue:

> This battle now moves to its next stage, where many thousands who have campaigned against this incinerator will look for something in the planning, political and judicial process that can deliver them true justice. (*ibid.*)

However, the state's response to CHASE, which included the granting of a limited degree of access through the oral hearing, followed by political closure through the overriding of An Bord Pleanála's findings, represents a reversal of fortunes for the CHASE campaign, and indicates the limitations of the oral hearing approach for public campaigns.

The CHASE demonstrates how creating 'increased access' (*ibid.*) in their ability to create 'alignments' with existing environmental activists in the Cork area limited support from allies such as An Bord Pleanála's inspector provided assistance, nevertheless, 'the strength of state' (Tarrow, 1984, p. 89) was the decisive factor, as the state overruled the Bord's inspector, in the interest of pursuing its wider policy aims.

In June 2011, the plans for an incinerator in Cork Harbour were finally rejected. The details of this can be seen in the following article from the *Irish Times:*

> An Bord Pleanála has refused planning permission to Indaver Ireland for a proposed €150 million twin incinerator project in Cork harbour. An Bord Pleanála signed off on the decision yesterday for the proposal for the Ringaskiddy site and communicated the decision to the various parties today. Indaver Ireland had originally applied for planning permission for a 100,000 tonne hazardous waste incinerator, a 140,000 tonnes municipal waste incinerator and a transfer station. In January 2010, Bord Pleánala refused planning for the municipal waste incinerator but said it was considering granting planning for the hazardous waste incinerator and the transfer station. It said Indaver should make some changes to its environmental impact statement to address concerns over flooding, coastal erosion and revised emissions. In May 2010, Indaver reduced the size of the buildings by 40 per cent while retaining the same capacity while it also made changes so water would be recycled rather than discharged. Environmental group Chase has campaigned against the project saying the proposed site is unsuitable due to risk of flooding and erosion.
>
> During a four-week oral hearing in May and June 2009, the group also questioned the need for the facility and expressed concern about its impact on public health in the harbour area. In a statement Indaver said it is "disappointed but not discouraged" by the decision as the reasons given by an Bord Pleánala can be addressed. "Our proposal aims to address the deficit of waste infrastructure in the Cork region, following today's decision, that deficit still exists and it simply won't go away.
>
> "It is no longer responsible, credible or efficient to simply export or landfill our waste. If we are serious about protecting 50,000 jobs in the pharmaceutical sector, then Ireland must provide the necessary infrastructure for its own waste. "The choice now for all of us is to decide whether we continue to ignore the obvious deficit or finally deliver solutions. Indaver remains committed to being part of the solution," the company added. (Cassidy & Roche, 2011)

3.2.4. Meath: The No Incineration Alliance

In the eastern region of the country, a major incineration plant was also planned for a site at Carranstown, Duleek, County Meath, to deal with waste from the Leinster region of Ireland. A local group, known as the 'No Incineration Alliance' was formed soon after plans for the Duleek plant were announced. The Meath group were able to establish links with GSE whose campaign was by then underway. This was an attempt to establish 'influential alliances' and networks with existing anti-incinerator groups in Ireland. Both groups provided mutual support, with members of both groups attending each other's public meetings. GSE's Conchúr O'Brádaigh spoke at meetings of the Meath group. The alliance ran a publicity campaign to highlight the danger of the dioxins contained in incinerator emissions. A series of public meetings was held, and a website 'no incineration.com' was established, along the lines GSE's own website, which was established at the onset of their campaign, providing material, evidence and information about the alliance's campaign.

The alliance's main strategy followed the established pattern of an oral hearing, which commenced in October 2002. The campaign proposed to address the relevant planning, legal and EPA related issues through its campaign. In an overview of the Alliance's concerns presented to the hearing, health risks/fears were emphasised for those living in the vicinity of and in the wider area of the proposed Duleek plant. The concerns were based on the negative experience of incinerators in Europe and the United States. In addition to the risk to humans posed by an incinerator, concerns for wider environmental impacts were also raised in the appeal document. The Duleek site was also rejected, as it was 'fundamentally inappropriate for the proposed development' (Planning Appeal of No Incineration Alliance, 21 October 2002). NIMBYist concerns were raised in the appeal about visual intrusion, as well as damage to agriculture and tourism. The Meath Campaign exploited these concerns, framing their challenge in an appeals document. The issues included in this framing process included unsuitable location, as well as negative visual impacts. The NIA's ability to exploit NIMBY concerns about local agricultural and tourism impacts provided their campaigns with a broader appeal, in the absence of the wider mobilisation achieved in urban centres such as Galway and Cork.

The appeal document goes on to question the compatibility of an incinerator plant with an area zoned for primarily agricultural usage. This approach to framing the issue in the appeal document represented an attempt to highlight the impact dioxins had on agriculture, which was still the primary economic activity in the area. By linking the Duleek site with local Heritage areas, the NIA were exploiting a political opportunity to frame their campaign in a way that included as wide a range of issues as possible. It suggests that

> The applicants have failed to make a sustainable case for the location of this development at this particular site. (No Incineration Alliance Planning Appeal Document, 21 October 2002, p. 7)

Although the Duleek site was not set out in any special area of conservation, it is close to a proposed National Heritage Area of wetlands commons. The appeal document argued that an incinerator would disrupt the existing hydrology of the area.

The alliance put forward their concerns about the destination of such hazardous waste and its by-products and claimed the incinerator plans materially contravened the provisions of the National Waste Management Plan in the following manner:

> Visual Intrusion: The issue of visual intrusion must ... be examined in the statutory landscape designation that relate to the area in which the subject site is located. (No Incineration Appeal Document, 21 October, 2002, pp. 9–10)

The alliance also included the tourist industry in their framing process, exploiting the opportunity it offered to extend their campaign's remit. The appeal document goes on to point out the fact that the subject site is in the Boyne Valley, an area that has been designated as a World Heritage Site with immense archaeological importance, and includes areas of visual quality.

> Impact on Tourism and Heritage: 'The landscape in which the subject is to be located is of international significance in terms of history, archaeology and landscape. For example, Newgrange (which is one to the three top tourist destinations in Ireland) is just 10 km away from the subject site'. (No Incineration Appeal Document, 21 October 2002, p. 12)

By highlighting the risks to local water tables, the NIA were extending their framing process to include local ecosystems. This provided a further opportunity to create consensus around grievance.

> Impact on Ground Water: 'The (North Eastern) Health Board voted that the subject site is located on a limestone reserve and raised questions regarding the likely contamination of the underlying aquafier given the nature of the proposal ... The fact that the Health Board's serious concerns went unaddressed is a matter of considerable concern. The issue of ground water contamination is a very serious one with implications for the health of water consumers within the general vicinity of the site'. (No Incineration Appeal Document, 21 October 2002, p. 12).

The alliance suggested that the proposed incinerator site's 'proximity to water-table sites', as well as concerns for surrounding flora and fauna, 'leaves a considerable cloud of doubt regarding the appropriateness of the subject site's location' (*ibid.*). One post-Celtic Tiger boom issue that could be seen to have both ecological and human impacts was the increased traffic that an incinerator would create. The document highlighted the issue as follows:

> Traffic: it is clear that ... the current proposal would have an impact on the existing pattern and amount of traffic in the vicinity of the subject site. (No Incineration Appeal Document, 21 October 2002, p. 13).

The alliance was concerned about an increase in delays, congestion and 'possibly hazard' on the existing road network, which would add to the negative impact imposed by an incinerator plant on the site. Their framing process also included direct appeals to local 'post-boom' concerns about property prices.

> Property Devaluation: it is our contention, that the proposed development will negatively affect on existing property prices due to its scale and nature ... In this regard we would submit that, notwithstanding the quantifiable or tangible impacts such as visual intrusion and noise, the very nature of the proposal would in itself lead to a significant devaluation in property. (No Incineration Appeal Document, 21 October 2002, p. 13)

Here, the alliance framed their arguments, suggesting that an incinerator would be a negative feature in the wider public's perceptions of the area that can, in itself, reduce property prices.

The appeal document went on to highlight further inadequacies with the Environmental Impact Statement, in the actual identification of the site, as well as in a failure to adequately consider alternative sites. A failure to consider the cumulative impact of the plan, on resident's property prices or the surrounding environment is again highlighted.

One of the ways the NIA tried to extend their framing process was by highlighting the EU's own waste hierarchy. The fifth and sixth grounds for Appeal were based on the plan's contravening of the EU's own waste strategy. By proposing an incinerator, the Alliance argued that Indaver were going against the EU's own 'most favoured' options for waste management, which starts with minimisation and reuse, has energy recovery and disposal, and incineration only as one of the 'least favoured' options. The alliance argued that the lack of plans for waste separation at the collection point meant that all waste would be incinerated, leading to increased levels of toxic emissions. The appeal document sets out the National Targets for recycling, and claims that 'it would be immoral to burn resources which could be used if processed and recycled' (*ibid.*).

The document goes on to argue that incineration is a 'contravention of principles of sustainable development' (*ibid.*), citing the UNCED Earth Summit in Rio, 1992, which claimed that sustainability should be a key goal of our society. The alliance argued that incineration emissions contribute to greenhouse gases, and should be avoided, claiming there was a need 'for the developed world to take waste minimisation and sustainable development seriously' (*ibid.*).

The framing process included many of the accusations that had been levelled at the incineration industry in the past. The Alliance claimed that incineration was an 'outdated' technology, and has developed primarily for nations without enough landmass to provide landfill sites, such as Japan or Holland. They argued that the excessive costs of incineration, in addition to the negative effects on health and agriculture, make sustainable alternatives such as recycling a more suitable option for waste management in the northeast of Ireland. The No Incineration Alliance appeal document concluded with further grounds for appeal based on precedent decisions that denied permission for other projects in the Boyne Valley region, due to its sensitive location and value as a heritage site. The document ended by calling on An Bord Pleanála to overturn the decision of the County Council Planning Authority and refuse permission for Indaver's plant.

The group again tried to present an argument based on scientific evidence. However, despite the points contained in the Alliance's appeal, and its ongoing campaign of protest that was maintained throughout the hearing, Meath County Council granted planning permission for the country's first waste incinerator at the Meath site on 31 July 2003. The decision was granted on the condition that waste materials could only come from the North East of Ireland.

The campaign in Cork, led by CHASE, was focused on the local impact of incineration, and the movement had a primary concern with nearby communities. While having the benefit of being able to draw on experienced environmental campaigners from other campaigns, the NIA campaign were unable to mobilise networks and alliances with like minded groups, beyond an initial level remaining focused on localised concerns. While NIA had attempted to frame the issue in a broad manner, this framing process lacked the degree of focus that both GSE and CHASE projected on to the health risks issue. Strategically, NIA never politicised their campaign, and were therefore unable to regain wider access to the political opportunity structure. Ultimately, the CHASE and NIA campaigns against incineration demonstrated the importance of a specific framing process, which could clearly highlight a societal grievance such as health risks. However, neither campaign was able to achieve the access to political structures necessary to create changes in the state's waste policy.

We have shed light on the NIA's ability to utilise 'influential allies' (*ibid.*) through their oral hearing on the Meath plant. 'Alignments' were also created with local communities, in relation to the potential for visual and environmental degradation in the area. The NIA also attempted to exploit the state's 'unstable alignments' (*ibid.*) with supranational institutions such as the UN and the EU, when it claimed that incineration ran counter to the state's responsibilities in relation to sustainable development.

3.2.5. Conclusion: The Emergence of Campaigns against Toxicity and Waste: 1970s–2004

This chapter has explored Irish multinational led industrialisation and economic growth between 1958 and the 1990s, while examining the emergence of environmental campaigns during two main phases of this growth. These two phases included the growth of opposition to multinational led development in the 1970s at Raybestos Manhattan, the convergence of rural communities and business interests in the 1980s, during the Hanrahan dispute, and the cross-border campaign undertaken in Derry at the DuPont Plant. While these campaigns emphasised the health-risks posed by toxic industries, the economic growth of the 1990s posed a new threat, as rising consumption led to a waste management crisis. The 1990s also saw the emergence of a legislative framework to deal with waste management issues. This framework is dealt with in Chapter 4.

The 1990s gave rise to disputes about landfills, as local communities began to resist any further reliance on landfill as the primary local solution to waste. As the waste issue came to prominence on the island of Ireland, the 2001 study on waste management strategies provided 'all-island' solutions to the problem. This study is critical of the use of incineration as a short-term solution to the problems of waste management on the island.

The campaigns detailed throughout this chapter are indicative of an 'issue-history' (Szasz, 1994, p. 162) pointing towards a 'social process' (*ibid.*) that links the different aspects of the Irish environmental movement in its various forms. This issue history approach to the creation of a path dependent series of campaigns allows for a greater understanding of the context of GSE's case. By taking an issue-history approach, further understanding of the two phases of environmental campaigning in Ireland can be presented, in a pre and post Celtic Tiger context. As the POS evolved on various levels during this period, external political factors, such as economic growth, wider North-South cooperation, EU legislation and the formation of the EPA all influence the events surrounding these cases.

Internal factors, such as the introduction of interest led experts such as Paul Connett or James Neufield established the 'influential allies' necessary to provide the data to support public campaigns networks were established through these experts with foreign campaigns, while eventually, networks began to be forged between local Irish campaigns These factors created further opportunities for movements such as GSE to link with and build on, as part of their own campaign. These various movements shared certain features, and drew from similar pools of internal and external resources, as an anti-multinational and NIMBY frame in the 1970s and 1980s gave way to concerns about the negative impacts of the economic boom in the 'waste management' frame of the 1990s and 2000. GSE's own campaigns were able to avail of these internal and external resources, as the shifting dynamic of political opportunity structures evolved, and 'influential allies' such as Dr. Paul Connett, who had been involved in the DuPont/Derry dispute were available. The volatile POS also provided 'unstable alignments' (Tarrow, 1994, pp. 85–89) as the state's own legislative framework on waste issues remained concentrated on profitability rather than sustainability. The following chapter explores the emergence of that legislative framework, as the state began to grapple with waste issues.

The campaigns against incineration in Cork and Meath demonstrate the varying degrees of success which can be achieved by movements utilising collective action frames and exploiting political opportunities to mobilise around grievances. While the CHASE campaign provided an emotive frame, as women and children attempted to highlight the potential health risks posed by incineration, their success, which included the ruling of An Bord Pleanála's senior advisor against Indaver's proposed plant, was met with an exclusive state response, which stated that policy superseded local concerns. Meath's NIA framed the issue in a broad manner, but this may have been too diffuse, and the NIA never mobilised the 'influential allies' (Tarrow, 1998, p. 79) necessary to allow for a successful exploitation of the 'dimensions of opportunity' (*ibid.*) open to the group. One group that combined the framing of risk with legal challenges were the various anti-landfill campaigns, which could exploit the political opportunity presented by EU waste legislation. This combination of successful framing of the issues of contention, combined with the utilisation of EU directives as the basis for legal challenges, represented some success for community groups opposed to the state's waste policy.

CHAPTER 4

SUSTAINABLE DEVELOPMENT AND THE STATE'S WASTE POLICY FRAMEWORK

4.1. INTRODUCTION: THE STATE AND WASTE

This chapter explores the state's response to the waste crisis (see also McDonagh, Varley, & Shortall, 2009). The conceptual basis for key turning points in the state's waste management policy is located within the parameters of an EM approach. An outline of eco-modern and sustainable thinking is provided in the chapter, as the state's policy shift on waste, from a reliance on landfill to a strategy informed by the EU's waste hierarchy would provide many of the political opportunities for GSE, and their political allies, to exploit.

As these changes in policy show, the groups attempted to frame this policy response within an understanding of 'stable' and 'volatile' elements of a shifting or dynamic political opportunity structure (Gamson & Meyer, 1996, p. 4).

The relevance of this policy framework to a study of an environmental campaign is due to the fact that this policy agenda becomes part of the external political structures that movements attempt to gain access to. Rootes (1994, p. 1) argues that the conditions surrounding institutional arrangements are 'contingent and temporary', creating access to political opportunities for challengers when these institutional arrangements are changed. In the Irish case, the state has attempted to introduce a more 'stable' arrangement in relation to waste management, through institutional and policy responses. The response of environmental challengers such as GSE can be seen as injecting a degree of volatility to the POS surrounding the waste issue. Blowers (1987, p. 292) defines the state's response to economic change as less ideological and more structural; incremental institutional change as represented by the state's evolving waste policy agenda from the establishment of the EPA, through the Waste Management Acts and Policy Statements of the 1990s. Therefore, the evolution and institutionalisation of a waste management policy provided environmental movements such as GSE with what Rawcliffe (1998, p. 57) has described as the 'points of leverage' to the POS surrounding the state's waste management policy.

However, it is arguable that the Irish state's emphasis on 'top-down' science and technology-driven solutions led to a focus on the inclusion of incineration in their regional waste plans. This was despite the fact that incineration was the second least-favoured option of the EU's waste hierarchy. This may have been a direct knee-jerk response to the EU waste management directives, which called for Ireland to reduce landfilling. The waste crisis should be understood in the context of the increase in generated waste between 1998 and 2001, which was reported at an average of 10.5% per annum (EPA, 2002, p. 13). The EU's waste 'hierarchy' placed recycling and reuse as the most favoured option, and landfill as the least. The hierarchy takes the following structure:

(i) Prevention
(ii) Reduction and Reuse
(iii) Recycling and Composting
(iv) Recovery
(v) Incineration
(vi) Landfill

(*Source*: EU Waste Management Framework Directive 91/156/EEC)

In terms of the key concept underlying preventative environmental management is that technologies should be as efficient as possible in their use of environmental services (*ibid.*). EM is an alternative conceptual approach to what were previously seen as problems with the 'end-of-pipe' solutions. Any costs were shown to be beneficial, as they would outweigh the future costs of repairing the damage caused by pollution. This approach coincided with a new, consumer-driven demand for eco-friendly goods, which arose in the 1980s. At a policy level, EM has been used to help forge new and coordinated approaches to what were previously seen as problems facing the differing strata of bureaucracy. Previous inefficient approaches to these problems had led to a crisis in legitimacy and regulatory failure. As a result, the eco-modernist discourse has provided a fresh approach, or what Weale (1992) calls a 'new politics of pollution', aimed at harmonising previously competing actors with an improved societal arrangement as the reward.

This chapter charts the significant junctures that came to shape the policy response of the state to their EU directive obligations, while simultaneously attempting to respond to concerns which were raised by local communities. These concerns were, in the main, focused on the siting of existing landfills and proposed 'superdumps' or incineration plants. In order to address the demands of both the EU and the domestic environmental lobby that had grown since the 1970s, the state established the EPA in 1992. Still, the Irish government faced external pressures from the EU in regard to waste management directives. As the economy expanded, greater levels of waste were being generated, increasing by up to 89% (Forfás, 2001). Articles 5 and 6 of EU directive 91/156/EEC demands that:

> The community as a whole, and Member States individually, should aim towards self-sufficiency in waste disposal ... (including) appropriate measures to encourage rationalisation, of the collection, sorting and treatment of waste. (Fagan, 2003, p. 73)

Ecological activists argue that policy must be measured in relation to its impact on both local and global situations and that any action thereof must as a result keep these dual impacts in mind. If EM holds out the promise of a step forward from an existing ecological consciousness, and as a determining factor within active society rather than as a peripheral player on the fringes, it can only do so by confronting the dichotomy between these two apparently irreconcilable opposites. When the tendency towards policy leans towards technocrats, the possibilities for EM diminish, in direct proportion to its place in the inherent motivations of a consumerist-fuelled, ever-expanding society that demands this productivity despite the increased ecological damage this productivity inflicts.

EM has become the subject of criticism from many of the more radical environmental groupings, who fear it will be used as an ecological 'catch-all' policy, while in reality many real issues on the environmental agenda remain unresolved. One result of this debate has seen the spread of shifting ideologies within the movement, resulting in the 'light and 'dark' green camps being established. Within these new paradigms, an accommodation of environmental concerns must be identified from the wide range of actors involved, including those in the wider society sympathetic to environmentalists' ideas, and ultimately with the state agencies and corporate institutions who shape and act on environmental policy (Leonard, 1999, p. 2).

By 1996 a waste management crisis had arisen, due to the imminent closure of many of the country's landfill sites. This growing waste crisis led to debate and the drawing up of *The Waste Management Act* of 1996. While waste material had previously been seen as lacking any value, discussions about the need for high waste levels for incineration, or the value in recycling material, had led to some reassessment, due in part to the promotion of EM and eco-efficient practices, particularly at local level. Transportation costs, available waste infrastructure and type of waste materials play a major part in any overall evaluation of the waste management process. The competition for waste material in member states with both recycling and incineration has increased as waste levels rise in these countries. Licensing systems for waste also led to price increases, as is the case in Ireland. What distinguishes Ireland from its EU neighbours on this issue is the lack of either a recycling or incineration infrastructure, such was the reliance on landfill over the years. In countries where such infrastructures exist, waste material becomes commodifiable goods. However, as Ireland had been traditionally a semi-peripheral state outside of the European industrial core, its limited industrial growth before the 1990s had led to very limited infrastructures for dealing with waste, so landfilling became the prevalent practice.

4.2. WASTE MANAGEMENT POLICY

By creating policies to protect the environment and promote waste management practices, the state attempted to change the behaviour of the public, corporate and municipal sectors, all of which had become over-reliant on the use of landfills. In some ways, the policy agenda introduced by the state to deal with the demands of the EU or concerned communities could be described as 'a blunt instrument' (Roberts, 2004, p. 156) to force through regulations that dealt with obligations on the one hand, while maintaining the focus on economic growth on the other. Concerned communities would challenge the state's choice of incineration as a preferred option due to perceived health risks. These contradictions would emerge in the wake of the state's attempts to create a policy framework to deal with the waste issue. The technocratic centralisation of 'top-down' decision-making powers on waste management itself created opportunities for challenges to incineration and the need for local politicians to support 'bottom up' pressure. And by omitting any discussion of the potential health risks posed by incineration, the state was left vulnerable to the challenger's contention that incinerators in other states had created health problems in their vicinities.

4.3. THE POLITICAL OPPORTUNITY STRUCTURE OF A DYNAMIC POLICY DISCOURSE: SUSTAINABLE DEVELOPMENT AND ECOLOGICAL MODERNISATION[1]

The state had established the EPA in 1992 to deal with environmental problems in general. Throughout the 1990s a number of Acts and policy documents were created in response to the waste crisis. These policy initiatives included the *1996 Waste Management Act;* the *1998 'Waste Policy Statement',* the *EPA's 2000 Millennium Report,* and the *Regional Waste Plans,* incorporating Dublin, the South East, Galway, Limerick, the North East and the Midlands. All these areas were to include a municipal incinerator as part of their waste management plan. The establishment of the EPA was seen as a progressive initiative by the state. However, the EPA has also been criticised as not being independent from the influence of the industrialisation imperative of the state (Taylor, 2001, pp. 35–70).

The overriding emphasis of EU environmental policy is focused on a form of sustainable development that supports economic growth while attempting to minimise ecological degradation. For member states such as Ireland, EU directives on environmental issues must be enforced in national legislation. Economic objectives have, in turn, environmental protections built in at the planning stage, due to a practice known as 'Ecological Modernisation' (Weale, 1992). By attempting to include the cost of environmental protection EM policy attempts to regulate the process of pollution minimisation, as well as the costs. The state then promotes environmentally progressive business practices moving from 'end of pipe' solutions to a reorganisation of industrial and environmental policy now combined in overall sustainable practices throughout the manufacturing process. While the intent of EU environmental policy is clear, non-compliant or environmentally 'laggard' member-states continued to attract direct foreign investment from US multinationals, with EU directives on environmental regulation reduced to a secondary consideration.

The broader contemporary understanding of sustainable development as a discourse has its roots in the 1987 Brundtland Report for the United Nations World Commission on Environment and Development (WCED) (Dryzek, 1997, p. 126). The Brundtland Report, titled 'Our Common Future', combined previously separate areas of analysis around the plight of underdeveloped nations, population growth, the effects of war and injustice, as well as global environmental concerns, into a vision of a linked process of economic growth with environmental amelioration, peace and justice over a long-term period of stabilised global activity (WCED, 1987). The Brundtland Report of 1987 was the catalyst for introducing the concept of sustainable development and was successful, despite the difficulties presented by the fractious nature of this form of social planning. In a parallel with Janicke, Hajer outlines environmental policy regulations as a central feature of the current ecological debate. Institutional arrangements must be both proactive and reflexive, their effects must be tested in the public domain through debate and investigation. Hajer traces this shift in the ecological debate that has led to the emergence of EM as the dominant paradigm. In the 1990s, in regard to policy and institutional change, ecological planning became an aspect of OECD nations' response to the crisis of legitimacy and state regulatory failure.

The links between technological and scientific innovation, and the flexible nature of ecological policy responses, are an example of how current and future discourse between institutions, state agencies and environmentalists will shape ongoing environmental regulations. They are linked by a common origin of necessity, and together with an outward-looking dynamic, focused on the future. These ongoing impulses are rejuvenated by a continuing reflexivity that provides the opportunity for ecological policy to take on a greater significance, in the face of rapid technological changes and their spin-off ecological impacts. This can be achieved by an analysis of social forces, or in Hajer's words by 'providing the sign-post for action within these institutional practices' (Hajer, 1995, p. 264).

It is within this sphere that we can identify the trends that have led to a transformation of the environmental situation, leading to EM's current position as a policy determinant of once unbridgeable forces. This acceptance of a 'plurality of discourses' is reflected in EM's emphasis on open government with integrated and 'horizontal' levels of dialogue between various agencies. This participation in 'policy discourse' revolves around differing capacities within social structures that have influence on the shift towards new criteria such as anticipatory policy making, and an emphasis on an economic, efficient form of environmentalism. Central to this is the concept that nature is a resource in itself, necessary to the public good, making it difficult for polluters to escape their responsibilities to a public now aware of its right to a clean environment. Hajer sees a major role for environmental groups in future policy making, leaving early environmental radicalism behind and moving to what he describes as a 'reconsideration of existing participatory practices ... based on the existence of a comprehensive environmental problem. Ecological modernisation acknowledges new actors … and shows itself in an opening up of the existing policy making practices' (Hajer, 1995, pp. 28–29).

In acknowledging EM's benefits, and by emphasising the cost-efficiency of environmental approaches, ecological modern policy holds out the promise of an effective method of capturing the institutional and corporate policy maker's imagination, placing the ecological agenda in the corridors of power as well as in the capitalist boardroom, in a way which would have been inconceivable under previous green movement approaches which concentrate on the participation of civic protest groups. However, this pro-business acceptance of increased sustainable growth can put ecological policy makers at odds with 'deep green' environmental activists, many of whom feel that EM is a ploy to diffuse widespread ideological protest. It is a claim to which environmental policy lies open to, chiefly due to its bureaucratic tendencies, together with its acceptance of aspects of modern technological advances. Ultimately, it remains to be seen how effective any ecologically-driven policy will be, bearing in mind the ever increasing capacity for global environmental disaster, particularly in relation to global warming and the proliferation of nuclear arms in states which have little or no stability.

By 1996, in Ireland, a waste management crisis had arisen, due to the imminent closure of many of the country's landfill sites. This growing waste crisis led to debate and the drawing up of *The Waste Management Act* of 1996. While waste material had previously been seen as lacking any value, discussions about the need for high waste levels for incineration, or the value in recycling material, had led to some reassessment, due in part to the promotion of EM and eco-efficient practices, particularly at local level. An opening of the processes of dealing with waste materials at all levels has been undertaken as a result.

The frameworks for the introduction of sustainable development policy goals have been debated recently. Daugbjerg and Svendsen (2003) argue that sustainable eco-taxes must be 'feasible' (p. 76) in an economic context, by rising eco-credits to reduce resistance to green taxes from the industrial sector. Their arguments underscore the need to create an acceptable climate for supporting eco-taxes and pollution control compliance by those industries that cause pollution, 'in a real world context' (p. 77). Refunds for controllers of pollution would increase compliance for green tax regimes, something that has been pursued in Scandinavia (p. 80).

Lafferty and Hovden argue that environmental policy integration (EPI) is at the core of sustainable development (p. 1). They put forward a framework detailing the extent of the dimensions of EPI. Placing the integration of achieving environmental objectives as a core objective of sustainable development policies, EPI should enable the achievement of specific sustainable policy goals at all of the stages of policy making and implementation while 'minimising the contradictions' between the various aspects of policy making (2003, pp. 7–9).

Discussing the environmental transformations undertaken by the state, Dryzek, Downes, Scholosberg, Herens, and Hurold (2002) argue that environmental concerns have replaced economic growth and redistribution as the main concerns of the state. However, their study focuses on highly developed states such as Germany and Norway, for whom economic conservation could now take its place alongside 'economic and legitimation imperatives' (*ibid.*, p.659). The policy framework established by the Irish state could be seen as an attempt to reach the same transformation, through EM and sustainable development influenced initiatives.

The concept of sustainable development was defined by the EU Commission as 'meeting the needs of the present without compromising the ability of future generations to meet their own needs' (WCED, 1987, p. 7). Further attempts to define sustainable development in strategic terms were undertaken at the United Nations Conference on Environmental Development in Rio de Janerio in 1992, which became known as the 'Earth Summit'. The concept of sustainable development provided much of the focus for the Summit, and the findings contained in the resulting report, entitled 'Agenda 21,' discussed the increase in consumption in the developed nations, as well as the increased environmental degradation that has accompanied this over-consumption (WCED, 1987). Agenda 21 focuses on local grassroots or 'bottom-up' reactions to environmental issues.

In relation to waste more efficient means of production are achieved by reducing waste at source. EM promotes an understanding of how current environmental problems will become an increased financial burden in the future, making environmentally sound approaches to industrial planning and government policy a 'best practice'. In addition, increased environmental awareness amongst the public has created a demand for environmentally sound industries and policy frameworks. As a result, EM offers a chance to promote environmental economic structures that would prove to be sustainable.

4.3.1. The Environmental Protection Agency

Bodies concerned with industrial performance, such as the IDA and the trade unions were represented on the EPA's advisory committee, which led to criticism from environmental groups. The agency was deliberately constituted so as not to be anti-industry, or to be perceived as such by prospective multi-national investors (Taylor, 2001, pp. 68, 69). The EPA's pro-business interest was made clear at the outset:

> Lest there be any misunderstanding, we are not picking a group of environmentalists to review licenses or to carry out a whole host of tasks which are being provided for in this legislation. (Dáil Debates 1992 from Taylor, 2001, p. 69)

A prioritisation of economic requirements over all others was intended as the government was concerned that the EPA could become an environmental rather than industrial agency, as the independent but state-linked agencies of An Foras Forbatha (National Institute for Planning and Construction) and An Taisce (The National Trust) had become. The EPA's initial budget was set at 8 million pounds but by 1995 it was reduced to only 5 million, thus lessening its potential power. Another area of concern was a regulation that allowed the Minister to delay the implementation of EC Environmental Directives, with no chance for a public hearing on such decisions.

Essentially, the EPA represents an 'official' form of environmentalism one that is wedded to maintaining Ireland's 'climate for investment' (Leonard, 1988). Official environmentalism draws on forms of professional expertise that have their basis in economic planning and the sciences and that work through formal political processes. These processes include the following:

> public education and extension of centralised regulation which is concerned with the establishment and authorisation of centralised planning procedures, enforcement of environmental controls through legislation, and the establishment of centralised institutions for research and monitoring. (Tovey, 1992, p. 276)

While Ireland remains broadly supportive of EU membership and its subsequent funding dividend, Irish environmental policy is formulated primarily through the European commission in Strasbourg, with limited input from Irish EC commissioners. However, the implementation of this policy is then shaped to suit the neo-corporatist economic agenda promoted by the Irish state. In this way, Ireland takes on aspects of what has been described as an environmentally 'laggard' state (Weale, 2000, p. 466) taking a minimalist approach to environmental policy creation and implementation. In contrast, environmental 'leader' states, such as Germany and Holland, dominate EC environmental thinking. This has led to the

'Europeanisation' of Irish environmental policy (O'Hearn, 1999) (Flynn, 2003), a policy process,

> that has led to an over-reliance on environmental directives, and a loss of input and cohesive support amongst the Irish public for environmental legislation emanating from the EC. (*ibid.*)

One of the most significant aspects of the EPA's new responsibilities lay in its attempt to implement policies in line with the innovative concept of integrated pollution control (IPC). The central feature of IPC is based on the establishment of an overriding system of pollution containment, which links together the various features of the environment, such as land, water and air, and recognises their intertwined relationship. Scannell outlines the areas covered by the EPA's Licensing Regulations (1994), where a system of IPC would bring a diversity of categories under one central authority. These include; minerals and other materials, energy minerals fibers and glass, chemicals food and drink, wood, paper, textiles and lather and cement and waste (Scannell, 1996, p. 525).

While a large area of the environmental landscape was now under a focused and centralised legal structure through the EPA, the criteria that surrounded its establishment, such as the priority given to increasing Ireland's industrialisation as well as economic and political considerations, have left their imprint on the Act's legal framework. For instance, Section 52 of the EPA Act outlines how it should 'provide support and advisory services to local and public authorities in relation to the performance of their functions' (EPA Act section 52). This section of the act deals with the highly contentious area of local authority's environmental performance, yet it in no way specifies how such 'support and advisory services' (*ibid.*) should be provided. The controversial nature of how local authorities deal with the IDA, multinationals, and the agricultural sector is not legislated for, while equally important policy formats such as IPC are introduced without a clear indication of the overall basis or direction such IPC controls are intended to achieve (Leonard, L. 1999, pp. 64–65).

A central feature of IPC is the concept of BATNEEC or 'best available technology: not entailing excessive costs'. The EPA's own handbook on IPC states that 'the technology should be best at preventing pollution in the sense that it is procurable by the operator of the activity concerned'. Technology itself includes techniques and the use of techniques, such as training and maintenance. NEEC set out the balance between environmental benefit and financial cost (EPA guidelines on IPC, p. 3). The contradictions between the spirit of the EPA and its operation have been criticised by those who feel the EPA should be following the precautionary principles of the EU (Taylor, 2001, pp. 48–50).

Coming as it did into a relative void of environmental legislation and policy planning, the EPA Act should have been seen as an opportunity to introduce far-reaching and comprehensive environmental protections to the domestic political scene, in such a way as to implement already existing EC directives (*ibid.*).

Problems of equality of public access, in terms of format and agenda, as well as difficulties for the public in approaching a legalistic framework are areas that are left unanswered by the EPA legislation, and point to the act's structure having been built around the state's policy framework. Pollution prevention becomes, by extension, pollution management and even pollution toleration, to facilitate industrial expansion.

If these components are implemented by agencies such as the EPA, or advocated in policy by the Irish Green Party, then the projected outcome for Irish society should be built around a cohesive equality in that society that sees development maintained with an environmental balance. The EPA has as one if its objectives the aim of achieving sustainable development, yet the specific aspects of the Act make it possible for industries to avoid environmental responsibilities (Leonard, 1999, p. 75).

Following on from the EPA and, 1996 Waste Management Act the state's response to the bourgeoning waste management crisis had its origins in the *1996 Waste Management Act*. This was followed by other benchmarks in waste management policy such as the 1998 Waste Policy Statement *Changing Our Ways* and *The 2000 EPA Millennium Report*, all of which culminated in the Regional Waste Plans. The following sections will examine these milestones in the creation of the state's waste policy position, by examining each of the aspects of this policy, and then examining *The Connacht Waste Plan* (CWP), which contained the proposals for incineration that in turn and we will see how they led to the campaign by GSE.

Initial EC law drawn up in 1975 through the Council Directive on Waste (75/442/EEC) left individual member states to interpret and give definition to waste in the context of their indigenous industries. This open interpretation led to some confusion about waste definition, and led to the drafting of Regulation 259/93 in 1984, which set out a common definition of waste. A subsequent Directive on Waste in 1993 led to extensive restructuring of the legislation, with a clear definition of waste as something that has been 'disregarded'. A 'List of Waste' was published in late 1993 as part of Council Directives 75/1442 EEC on Waste (94/3/EEC).

The Waste Management Act of 1996 includes a number of categories of waste. These include:

> Residues of production or consumption, expired products, material which have undergone any mishap, contaminated material, unusable parts, residues of industry of pollution abatement or extraction processes, and production for which the holder has no further use. (*EC Waste Management Act* (3/1) 1996)

In the context of incineration, these waste materials may have 'no further use' but can create dioxins when burned, according to GSE.

The Act did not apply to:

> emissions, effluent, sea waste or radioactive substance (ibid). The EPA established a registration system to cover waste management processes in Ireland, in accordance with the Act. However, the EPA did not establish as rigorous system as the Act would have allowed for. This registration system is much less onerous than the provisions on waste licenses or local authority permits. (Laurence, 2000, p. 55)

As the Act did not cover emissions, any pollution from incineration would have to be covered by existing IPC pollution controls. One reason why the waste crisis has developed in Ireland is that growing urbanisation had brought populations within the vicinity of once-remote dump-sites. A result of this increased proximity to landfills is an increase in community resistance to any further landfilling in these areas. The waste crisis has been characterised as reflecting a lack of overall consensus on the issue, due to the:

> lack of environmentally suitable waste disposal locations, poor quality management of many landfills, High Court Orders leading to the closure of some existing landfills and public resistance to new landfills and to proposed incinerators (O'Sullivan, 2000, p. 20; EPA, 2000).

4.3.2. The Waste Management Act 1996

By 2000, Ireland had implemented the EC Community Strategy for Waste Management that included an overall hierarchical structure for dealing with waste issues. Emphasis was placed in the strategy primarily on waste prevention, recycling and reuse. The 1999 Landfill Directive (1999/31/EC) limited the use of landfills, leaving the Irish government with no option but to search for alternatives. Local Authorities had already committed to this switch to alternative waste management process by the *1996 Waste Management Act*. However, the local authorities did not promote such alternative processes such as recycling:

> Despite Government and EU policies which support waste prevention, minimisation, reuse and recycling very few of the waste management plans produced by local

> authorities since 1997 give priority to these waste management strategies. Instead, most of these plans dismiss all options except landfilling or incineration...and there has been little meaningful public involvement or participation in their preparation. (O'Sullivan, 2000, p. 23)

This 'top-down' approach to the regulation and disposal of waste led to an emphasis on two primary waste management options, landfilling or incineration. There was less emphasis on prevention, reduction and recycling, the most favoured options of the EU, and community groups like GSE. In the Dublin region, targets for this objective between 1994 and 2004 were set at between '59% for recycling, 25% for thermal treatment, (incineration) and 16% for landfill' (Swords, 2002, p. 52). This target was to be achieved despite the 1999 figure of up to 95% of household waste being land filled in Ireland, a figure matched only in Russia among our European neighbours. Government policy on waste management was outlined in the document '*Waste Management-Changing Our Ways*' in 1998, while Regional Waste Plans were drafted at this time. The next section of this chapter examines these aspects of the government's waste management plans in detail.

The formation of the EPA in 1992 had provided an agency to provide a structured response to European Commission directives on environmental issues, within the pro-business demands of the corporatist partners and the IDA. As waste management became an increasingly problematic issue, a Waste Management Act (WMA) was enacted in 1996. The WMA provides an overall public authority statutory response to the challenges presented by the waste issue involving the EPA and local authorities. The role of the Minister for the Environment, Local Authorities and the EPA were redefined to incorporate them into a common framework for waste management policy in Ireland. The Act presents a regulatory framework for improved environmental standards and performances in relation to the practices of prevention, minimisation and recycling, in accordance with EU requirements. The Act was also devised with a structured development of a waste industry in mind, one that would respond to the new technological demands of the EC's directives on waste management, such as new methods of waste collection. The Act represents an attempt by the state to respond to the waste directives of the EU, through the establishment of a policy framework that incorporate the state, local authorities and the EPA.

The core provisions of the Act include penalties of up to £10 million in fines and liability for pollution clean-up costs. Agricultural, commercial or industrial obligations to prevent pollution from the design stage through the production process are dealt with in Section 28. The transportation, storage and disposal of waste are dealt with in Section 32. Section 39 deals with the licensing of disposal facilities and the prohibition of unlicensed disposal. Over time, illegal dumping was becoming a major issue, as increased dumping charges created an illegal dumping network.

The responsibilities of the relevant public authorities are also set out in the Act. The EPA was deemed responsible for waste recovery and disposal licensing, and for the management and control of hazardous waste. The IPC licensing system that was introduced by the EPA in 1992 has been maintained through *The 1996 Waste Management Act*. The IPC approach takes a broad approach to pollution control through licensing. IPC approaches focus on all forms of pollution, seeking controls on:

> environmentally optimal risk-reduction of pollutants with cost considerations imposed as constraint, implicitly or explicitly, on reduction effort. (Weale, 1992, p. 111)

The regional, city and county local authorities have undertaken the functions of the WMA. Local authority responsibilities include a planning process to deal with non-hazardous and commercial waste, domestic waste collection, recovery and disposal, and an overall supervision of waste-related activities in their areas. The Act places overall power for waste management in the hands of the Minister for the Environment and Local Government. Ministerial powers are specified as follows:

> The Minister will be empowered to assist and support waste prevention and recovery initiatives, and to make regulations to promote waste prevention, minimisation and recovery. The Minister will also be required to promulgate a programme of in-house waste management to be followed by the public authorities. (WMA, Section 59, 1996)

The general provisions of the Act provide for new powers for the EPA and local authorities, which are aimed at facilitating legislation enforcement and monitoring. Sections 14 and 15 dealt with the powers of inspection and obligations towards monitoring. A framework for waste management planning is then outlined in the Act. The EPA is charged with establishing an overall hazardous waste plan in Section 26, offering recommendations on such waste to the local authorities. The local authorities must prepare individual or joint waste plans for their region. These regional waste plans are required to follow EU directives on waste management, and as such should incorporate prevention and recovery methods. Procedures for public consultation were also incorporated into the regional waste plan process, but only submissions were accepted as the regional plans were introduced, something GSE seized on.

The Act goes on, in Section 28, to deal with regulations on waste prevention and recovery. The obligations of all waste producers such as commercial producers and consumers in industry or agricultural are outlined. Obligations are imposed to minimise or prevent waste production, through production or manufacturing processes. Sections 29 and 30 provide the minister with regulatory powers for waste recovery. Environmental targets on waste recovery are then set by local authorities, in line with EU directives.

Local authorities are charged with the allocation of commercial permits for waste collection, as this sector increasingly is privatised. Private operators have emerged, as the sector has become regulated. However, illegal dumping remains an issue, as sites for dumping become scarcer. Laws for the separation of waste for recycling have been created. In regard to waste recovery and disposal, the local authorities have a duty to provide the necessary infrastructure. The licensing of waste facilities is covered in Section 39 in the Act, and is under the control of the EPA. Waste licenses must be issued through the BATNEEC system, to reduce or prevent emissions or other pollution.

Although there is some scope for public consultations or objections through the Act, it is limited. The EPA is the agency that will adjudicate in regard to such hearings, with appeals being referred to the High Court. Section 54 removes waste licensing from the planning process and is dealt with by the EPA. The Minister for the Environment retains power of exemption over waste licenses. Local Authorities are given power to enforce compliance with notices on regulation breaches, with any such action having priority over other regulatory provisions. Section 64 provides for an emission inventory, which is to be supervised by the EPA, to create an understanding of the need to reduce hazardous waste emissions. The area of planning remains contentious. All aspects of *The 1996 Waste Management Act* incorporate the 'polluter pays principle', which shifts the emphasis from end-of-pipe solutions to an incorporation of the concept that costs should be borne by the polluter. The Act's provisions came into effect from 1 July 1996.

4.3.3. Waste Policy Statement 1998: 'Changing Our Ways'

By 1998, as the provisions of the Act began to impact on social processes, the government introduced a wide-ranging policy statement about waste management titled '*Changing Our Ways*'. The statement reflected a common concern in government circles that waste generation was continuing to increase, while there was a lack of progress in implementing alternatives to landfilling, which was deemed to be an unsophisticated approach in its current form, according to the findings from the European Environmental Agency's Report 'Europe's Environment: The Second Assessment' in the Spring of 1998. By September of that year, the then Minister for the Environment, Noel Dempsey, had launched the '*Changing Our Ways*' policy statement. The statement commences with a recognition of the fact that

> There is an urgent need, in line with government policy and the new framework of the Waste Management Act, 1996, to modernise waste management practice and secure the provision of environmentally efficient infrastructure. (*Changing Our Ways*, 1998, p. 1)

As the Minister placed his own stamp on waste policy, tensions between local and national government were reflected in the criticisms of the policy statement aimed at local authorities. The statement critiques the existing waste services that had been operated by the local authorities for not introducing innovative management or technological solutions to existing waste problems. This had led to both private and public sector perceptions of local authority waste management being negative, creating a source of tension on the issue. In response to this, the policy statement sets out areas of focus for local authorities, to improve these perceptions. The local authority focus should be on household, commercial and industrial waste, with 'good waste practices' being introduced, as the authorities exert influence over the waste sector.

Local authorities are charged with taking 'a pivotal role' to effect these changes. The policy statement is presented as a continuation from other strategy documents, such as '*Recycling for Ireland*' (July 1994) and '*Sustainable Development: A Strategy for Ireland*' (April 1997). The 1998 policy statement's purpose was the provision of an overall national framework for both the local authorities and an emerging private waste industry to operate 'with confidence'. The main changes introduced through this framework include an infrastructure to cope with the regulation of the industry, and the licensing of waste management processes such as incineration, landfilling, and recycling. The transition from policy to the adoption and implementation of waste management plans is also recognised as in area in need of attention

The policy statement speaks of the need to 'address this challenge with vision and vigour' so as 'to secure radical change' and of 'the need to be ambitious' in relation to the implementation of waste policy (*Changing Our Ways*, 1998, p. 2). There is also a recognition that Ireland needs to go 'beyond a basic compliance with national and EU legislation, incorporating best practice and resource efficiency in economic sectors' (*ibid.*). The statement also reinforces the importance of the polluter pays principle as 'fundamental' to waste management efficiency, while also emphasising the importance of economies of scale that take into account the viabilities of waste strategies.

A reduction of the over-reliance on landfill is highlighted in the statement. According to the *National Waste Database (NWD) Report* of 1996, over 2 million tonnes of waste were landfilled in 1995. As such, the statement outlines the primary importance for local authorities in securing a reduction of the over-reliance on landfilling for the disposal of waste. Many of the smaller landfill sites are in the process of closing as they reach the end of their licenses, while others are nearing their capacity levels. According to the Waste Management Act 1996, the EPA will close any landfill that has not been upgraded to meet new environmental standards. This policy initiative will also lead to more efficient private landfills around the country.

Local opposition to landfills is recognised in the statement. 'Nuisance factors' are listed as including 'odours, dust, litter, noise, rodents, insects and general unsightliness' (*Changing Our Ways,* 1998, p. 4). The dangers of soil, water and atmospheric pollution posed by landfilling are highlighted. The limited nature of landfilling waste and its inhibiting of recovery and recycling options, are seen as contributing factors to the waste management problems faced by local authorities. In response government policy requires a huge undertaking, which includes diverting of 50% of household waste from landfill over a 15-year period to 2013, to meet the requirements of EU legislation which included a requirement for each member state to reduce landfill levels from national waste strategies within 15 years. The diversion of waste will incorporate two main areas, domestic and industrial. The diversion of construction and demolition waste is one area targeted for change, as most of the current construction waste landfilled is reusable. As construction waste has a high volume to weight ratio, this change could make an immediate beneficial contribution to the reduction effort. The construction industry is charged with implementing these changes, through its obligations under the 1996 Waste Management Act.

The waste licensing system introduced by the EPA through the 1996 Act will lead to a focus on good design and operation of remaining landfills, according to the statement. An aftercare system will also be operated by the Agency after landfill closure. As a result of the proposal changes to waste management, the statement predicts a future where financial costs associated with waste management increase significantly, as higher environmental standards are introduced. Reliance on landfill should be significantly reduced, while a waste recovery and recycling infrastructure are put in place. While such measures should increase efficiency, the costs of creating a waste management infrastructure will make any profitability a long-term aspiration.

The future of waste management is incorporated into the parameters of the 1996 WMA, in addition to a 'flexible legislative framework' (*ibid.*). The role of the local authorities in waste management, as well as 'a strong emphasis on public consultation' (*ibid.*), is presented as key components for waste management strategies. The process of planning for the future should incorporate the type of neo-corporatist arrangements envisaged under Agenda 21:

> A new approach to environmental management, involving constructive cooperation with local communities and neighbouring local authorities, and utilising the potential of the private sector to contribute in the delivery of public services. (*ibid.*)

A regional approach to waste management has been encouraged to provide for a wider service with an efficient infrastructure. The regional approach facilitates an 'economy of scale' framework for the local authorities to cooperate in providing viable planning for 'integrated and innovative waste management solutions' (*ibid.*). Local authorities are provided with the regional strategic policy that in the West of Ireland involved all the Connacht counties contained in an overall regional waste plan. This regional approach was intended to be an inclusive one, incorporating the local authorities, sectional interests, in addition to 'effective public consultation' (*ibid.*). The timescale for this process was set between December 1998 and June 1999. The creation of local authority and private sector partnerships is also seen as a positive future strategy, highlighting an ideological shift from state to market-based solutions.

The increased participation of the private sector in waste management is something that the policy statement encourages, to facilitate an increase in capital expenditure and investment, technological expertise, marketplace knowledge, combined with the positive effect of freeing up local authority staff and finances for use in other services. The statement goes on to explore the development of a market infrastructure for recovery and recycling, including partnership approaches and investment in waste facilities. This reflects the government's faith in a technocratic and market solution to the waste issue, using incineration technology in a 'market infrastructure'.

4.3.4. The 2000 EPA Millennium Report

According to the 2000 Millennium Report on Ireland's environment, nearly 80 million tonnes of waste were generated in Ireland in 1998, mostly from agricultural production. Two million tonnes of municipal waste were produced that year along with 91 million tonnes of industrial waste. This amounted to an over 100% rise in waste collected by local authorities in the years between 1984 and 1998. This period also witnessed an increase of 47% in the generation of industrial waste, combined with an increase of 13% in hazardous waste. In addition, the Report indicates that 2.7 million tonnes of construction and demolition waste were produced in 1998 alone, an amount which is projected to significantly increase due to the increase in construction taking place throughout Ireland.

The destination of this waste material has primarily been landfill sites, with as high as 91% of Irish municipal waste being landfilled in 1998. The Millennium Report charts recovery rates for packaging waste at 14.8%, although some progress has been made in this area during the intervening years. Industrial waste recovery improved between 1995 and 1998, rising from 12.4% to 26.6%. Manufacturing recovery rates during the same period improved from 30 percent to 51.4%, according to the Report, indicating some success for the EPA's emphasis on private sector partnership.

Priority issues included the minimisation of waste production, in addition to attempts to 'break the link between economic growth and waste production' (*EPA Millennium Report*, 2000, p. 58). This waste contained materials such as paper, glass, plastics, aluminium and other metals, textiles and organic waste. Only 9.0% of the waste generated in 1998 was recovered for recycling in contrast with the 50% recycling achieved in a pilot scheme in Galway in 2001. According to the EPA Report, Ireland differs from its EU neighbours by not having introduced incineration, a point that is seen to be at variance with the ethos of EU environmental policy:

> In the absence of incineration, landfill will continue to be the primary disposal route for household and commercial waste, despite the fact the EU policy considers landfill to be the least desirable waste management practice. (*EPA Millennium Report*, 1998, p. 58).

As recycling was the main alternative to incineration, according to community groups such as GSE, the affordability and infrastructure for this process would be crucial. Recovery rates for the purposes of recycling household and commercial waste increased from 7.8% in 1995 to 9.0% in 1998. A decrease from 4.3 to 3.2% in household paper waste collection in the same period was due to declining household paper waste collection during this period. However, household glass collection rates improved, from 7,900 tonnes in 1995 to 14,100 tonnes in 1998. Despite these successes, the EPA acknowledged that 'material waste recycling is a volatile and uncertain business' (*Millennium Report*, 1998, p. 58), due to fluctuation in the value of recycled waste, and to recycling rates remaining unpredictable, and market solutions remained problematic, due to prohibitive costs.

European Commission directives on packaging waste has called for an EU-wide approach to the recovery and recycling of glass bottles, aluminium cans and paper containers which make up packaging waste materials. Irish recovery targets were aimed at a 25% recovery rate for 2001, rising to 50–65% rate of recovery for 2005. The EPA Report goes on to highlight again the lack of incineration in the Irish case:

> owing to the absence of municipal waste incineration capacity in Ireland their targets have to be met by recycling alone. (*EPA Millennium Report*, 2000, p. 59)

Industrial waste grew from 6.2 million tonnes to 9.1 million tonnes from 1995 to 1998, an increase in 47%, most of which was produced by the manufacturing sector. Seventy three percent of this industrial waste was disposed of through landfill, with 27% recovered through the IPC licensing system. The 1998 figure for hazardous waste was 370,000 tonnes, up 13% (*ibid.*); 35.6% of this hazardous waste was disposed of or recovered on site with 11.5% disposed of off-site. A further 26.9% was exported. Construction waste accounted for 2.7 million tonnes, with 1.17 million tonnes of this being recovered for road fill (*ibid.*). All of these figures are indicative of the increased consumerism and construction waste, which was forcing the state to search for waste management alternatives.

As the creation of waste increases in Ireland, the EPA has decided on a number of options for the management of waste across the country. The preferred option is the prevention and minimisation of waste. This approach includes a wider social acceptance of a need to reduce waste, through consumer choice. The National Hazardous Waste Management Plan (1999) incorporates the IPC licensing system in an attempt at targeting producers, while providing financial and technical assistance towards waste reduction efforts. Legal frameworks target individual or corporate polluters. Cleaner production methods are also being promoted. While the EPA has had some success with pollution reduction through the implementation of the IPC system, others in the industrial sector are still increasing their waste.

The next best option for waste management is recovery and re-use, which according to the Millennium Report includes 'recycling and energy recovery' (incineration), with the added proviso that:

> when considering the net environmental impact, this needs to be balanced against the overall cost to the environment in relation to energy usage and pollution that might arise as a result of the recovery process itself. (*EPA Millennium Report*, 1998, p. 65)

One such area of pollution that would stem from the recovery process is the emissions from incineration. Incineration or 'thermal treatment,' is 'the subject of some debate' according to the Millennium Report. The absence of incineration in Ireland is in 'marked contrast to other European countries' (*EPA Millennium Report*, p. 67). The report indicates that the state's desire to increase efficiency in waste management as part of their eco-modern approach to concepts such as energy recovery.

The Report states that:

> it appears likely that Ireland will see the introduction of thermal treatment with energy recovery for municipal waste, particularly in large urban areas. (*EPA Millennium Report*, 2000, p. 66)

The pressure on existing landfill sites lies behind the attempt to introduce incineration in Ireland. Sites have become overburdened, with over 8.8 million tonnes consigned to landfill in 1998. Some 4.86 million tonnes of this went to privately operated landfills, with 3.94 million tonnes going to local authority sites. Sites that have the capacity to accept over 40,000 tonnes per annum exist in Galway, Ennis, Kilrush and Newcastle West in the West, in Cork, Waterford and Wexford in the South. Twelve such sites exist in Dublin and two each in Meath and Louth. These large-capacity sites correspond with the proposed sites for incinerators, as local authorities close landfills that have reached their intended levels.

The EPA forecast continued increase in waste production rates, due to the 'stubborn linkage' between economic growth and waste production. A link is established between the sustainability of our economy and the consumption and use for resources. The EPA has adapted a sustainable waste management approach in an attempt to minimise the impact of increased waste production. As the pressures on the environment increase through waste production, infrastructural development of roads, networks and increased construction, Ireland's accelerated expansion has come under further scrutiny from environmental groups.

The alteration of Irish waste management policy has resulted from 'the Europeanisation of Irish environmental policy' (Flynn, 1993, pp. 2, 3). The problems of waste management have led to what Flynn has described as the 'triumph of policy style over substance' in regard to environmental affairs (*ibid.*). Essentially, Irish policy has been forced through directive compliance to maintain the appearance of an environmentally conscious member-state. Despite what Flynn describes as the 'particularist nature' (2003, p. 2) of certain aspects of Irish domestic environmental opinion, Irish environmental policy follows the pattern of profound impact that occurred in every area of policy since EU membership began in 1973. However, some recent policy initiatives, such as the plastic shopping bag levy introduced in 2003, have been successful and popular environmental successes stories for the state.

4.3.5. The Connacht Waste Plan

The state's six regional waste plans in Dublin, the South East, Connacht, the Midlands the North East and the Mid West were conceived to provide local solutions to the waste crisis to allow 'an economy of scale' approach to the waste issue, which could deal with the dispersal population in rural areas. *The Connacht Draft Waste Management Plan* was prepared by MC O'Sullivan, in association with COWI incineration from Denmark in conjunction with Mary Murphy and Associates. It was devised as the Waste Management Plan for the Western counties of Galway, Mayo, Sligo, Leitrim and Roscommon, in addition to Galway City. The Plan was prepared in accordance with the Waste Management Act of 1996. Each of the relevant local authorities was presented with a Waste Management Strategy in the spring of 1999. This strategy recommended integrated approaches to waste management, such as recycling, biological and thermal treatment or incineration of waste, in addition to landfilling to environmental standards. The Draft Plan was put on public display at local authority offices for a two-month period before the Plan's adoption. The power to adopt the plan rested with the elected members of the local authorities.

The CWP attempts to shift local and individual local authority attempts to deal with waste to a regional approach that incorporates 'more innovative solutions to solid waste management' (CWP, 1999, p. 1). These solutions incorporate an 'economy of scale' approach, to allow for the demographic variations between sparsely populated rural areas, and more densely populated urban centres. This top-down integrationist approach to waste management is presented as facilitating an integrated approach to waste management in the regions, in addition to the beneficial partnerships that may be undertaken by the private sector and local authorities. Funding from the European commission was also forthcoming, under the European Regional Development Fund.

The plan begins with a review of the 1998 National Policy Statement on Waste. It reiterates government concern for a reduction in landfill dependency, through a transition to an integrated waste management system. In an overview of current waste management processes, a quantities table of non-agricultural waste produced in the Connacht Region is supplied. Included in these figures are:

133,115 tonnes per annum of household waste collected
53,162 tonnes per annum of commercial waste
201,510 tonnes of construction and demolition waste
46,672 tonnes of waste emanating from mines and quarries
6,219,679 tonnes per annum of agricultural waste
596,054 tonnes of waste from all other sources

Waste collection in Leitrim, Sligo and Roscommon was in the hands of the private sector while Galway and Mayo had local authority collections. Charges ranged from £40 (€50.79) to £120 (€152.32). Industrial and commercial waste was largely in the hands of private collectors.

Recycling is also undertaken through private operators, as well as through the bring-bank system. In 1999, only 3.5% of municipal waste was recycled, as opposed to the 15-year target of 35%. Materials collected for recycling are mainly metals, glass, paper and cardboard, as well as pallets, plastics and items such as batteries. Some of the problems of recycling include the fact that transport costs are prohibitive, and the lack of an existing, sustainable market. However, the increased costs associated with landfill and other treatment processes will in turn make recycling prices relatively more competitive.

In 1999, the Connacht Region had 12 working landfills that received approximately 200,000 tonnes of waste per annum. The Waste Plan proposed a reduction in the number of landfills to 6 by 2000, including the closure of the large Carrowbrowne site near Galway. Improved landfill sites that remain in operation would be self-funding due to the introduction of weighbridges and gate fees. A trend towards private collection is noted in the Plan. Waste processes in the region would also incorporate other processes, such as recycling and thermal treatment, according to the plan. This integrated approach should be undertaken through a best practicable environmental option (BPEO), leading to certain desired scenarios such as the achievement of 'maximum realistic recycling', the achievement of 'national and EU targets' by recycling and thermal treatment or the achievement of 'maximum landfill diversion through fastest possible implementation of recycling and thermal treatment' (CWP, 1999, pp. iv, v). By taking this approach the state hoped to solve the waste management crisis in an eco-modernist manner.

These scenarios incorporate likely future influencing factors such as possible change to national or EU targets on recycling. Projecting ahead to 2013 targets, 'assuming thermal treatment is in place by this time', set waste processing figures as follows; household waste: 44.6% recycling, 52.7% thermal treatment, and 2.7% landfill; Industrial waste: 27.0% recycling, 29.2% thermal, and 43.8% landfill, and construction and demolition waste: 79.8% recycling, 0% thermal, and 20.2% landfill. The Plan provides for one thermal treatment plant 'in Galway City Environs.' The targets aren't reflective of the EU's waste hierarchy, which favours recycling over this emphasis on thermal treatment.

Waste management policy is based on policy objectives such as waste minimisation. Waste minimisation will be included in the remit of appointed environmental awareness officers, who would

> lead new public education initiative in each local authority areas. IR£80,000 per annum to be allocated by each local authority to include for one environmental awareness officer (with administrative back up and budget for consumables) in each authority. (CWP, 1999, p. v)

However, one of the first duties of the newly appointed Environmental Officers was to visit incinerators in Germany. A regional industrial waste management officer has been appointed for the Connacht region, according to the CWP. The industrial waste management officer's role includes encouraging industries to minimise their waste output, while incorporating clean technology initiatives for industry, The RIWM officer would also be responsible for researching and presenting information on recycling markets and recycling industry opportunities.

The role of the local authorities would be to regulate waste collection, encourage waste minimisation from the source, and to enforce any subsequent waste collection regulations from the department of environment. Local authority housing complexes are to be incorporated into new waste management practices, as flat waste charges switch to user-related charges to encourage waste reduction. Concepts such as a pay-by-weight mechanism for domestic and commercial bins have already been introduced in some areas to enforce the polluter pays approach. Local authorities are also charged with enforcing new packaging regulation aimed at the industrial sector, which would encourage recycling rather than disposal.

In regard to local authority organisation, the authorities would establish waste planning and regulation units according to the waste plan, which would incorporate waste regulation officers, and a public education unit. Householders not covered by local authority collection service are to be monitored.

This inventory will then be used to extend waste collection to a figure of 80% of households in Connacht by 2005, and 100% by 2007, under new waste regulation by-laws. An overall consistency in waste regulation and charges will be implemented, while local authority staff are to be facilitated in the area of re-training, as new environmental regimes are monitored for progress. This overall method of implementing good waste management practices reflects the eco-efficiency inherent in an EM approach.

Waste recycling is collected from door-to-door, with source segregation in up to 50% of the region, with urban areas being targeted. New types of collection vehicles came into operation by 2000. The urban areas targeted for door-to-door recycling services include Galway City, Tuam, Ballinsloe, Oranmore and Athenry in county Galway; Ballina, Castlebar, Westport and Claremorris in County Mayo; Roscommon Town, Boyle and Castlerea in County Roscommon, as well as Carrick-on-Shannon in County Leitrim and Sligo Town. For rural areas, bring-bank services will be extended and upgraded. Towns and villages in rural areas of up to 200 people will be incorporated into the bring-bank network. Regional Materials Recovery Facilities are located in Sligo and Galway, the largest urban centres in Connacht.

A further 21 recycling stations will be situated in the major urban centres of Connacht, in the same towns where the door-to–door system operates, as well as Belmullet, County Mayo; Ballaghadereen and the parts of Athlone covered by County Roscommon; Tubbecurry in County Sligo, and Manor Hamilton and Ballinamore, County Leitrim. These centres should facilitate the recycling of the following materials: recyclable wastes such as glass and metals, bulky waste such as disused refrigerators, priority waste such as batteries, garden waste and light construction waste materials. A recycling scheme will also be started on the Aran Islands. This aspect of the regional plans can be seen as an attempt to move beyond a 'laggard' implementation position on waste management by the state.

The eco-modernist basis of the regional plans included plans for household and business segregation and organic waste collection to commence by 2003 in urban areas, extending out to a target number of 50% for rural areas. Commercial and construction industry waste will be incorporated into a source separation and recycling system, through the Materials Recycling Facilities in Galway and Sligo. Composting plants will also operate in the regions, as part of the CWP. Composting plants were opened in Galway City, Ballinasloe, Castlebar, Roscommon, Sligo and Carrick-on-Shannon. These plants would accept household waste as well as 'green' waste from recycling centres. Compost from these sites will be utilised by local authorities for park and landfill maintenance. Two biological treatment plants for kitchen waste have been located in Tuam and Sligo respectively. Treatment at these centres will be through composting, or by anaerobic digestion.

The most controversial aspect of the CWP was the inclusion of an option for a thermal treatment or incineration plant. The plant was to have an approximate capacity of 150,000 to 200,000 tonnes per annum, and would be cited in 'Galway city or its environs, serving the region' (CWP, 1999, p. viii). In other words a regional incinerator would take all of Connacht's 'non-reusable waste, and incinerate it at the region's largest population centre' (*ibid.*). The only siting criteria mentioned were factors such as industrial zoning and transportation needs. The proposed thermal treatment/incineration plant also had a potential to treat both municipal waste and dried sewage sludge in a combined process, an option that 'needs to be exploited' (*ibid.*) according to the Plan. Specific mention is made of the need for a 'public information campaign' on the inclusion of thermal treatment as part of the CWP. This process should be prior to a definite siting of the plant. According to the plan 'this process had already commenced in Galway City and Environs' (*ibid.*).

The plan highlights the need to improve the standards of operation at the interim landfill sites with immediate effect. These improvements can be met through 'better management practices and higher standards of operation' including:

> application of daily cover, control of pests and nuisances, improved fencing, improved visual appearance (litter control, better signage) infrastructural improvements (roads, compacting equipment) and weighbridge/data collection. (CWP, 1999, p. viii)

The recommendations for interim landfill improvements included upgrading and extending existing landfills to provide a countrywide capacity until the end of 2005. Remedial repairs are to be maintained until the new residual landfills and thermal treatment incinerator are in place in 2006. Transfers to outside landfills in neighbouring counties are recommended for the smaller counties, such as Leitrim. Medium to long term plans are to develop two regional residual landfill sites by 2006. One site proposed is in North Connacht, at Ballaghadereen/Charlestown on the Mayo-Roscommon border. Separate sites are proposed for Galway, including Newbridge, New Inn and Ballinahistle/Kilrickle. EPA license conditions will be applied to these sites. Inventory and aftercare will be provided for all closed sites.

The combined capital and operations costs needed for the implementation of the new waste strategy for the region are highlighted in the Plan. These costs would be met through private sector involvement, and through the implementation of collection costs and gate fees. Local authorities would have to meet the costs of public education and remediation of landfills. This would necessitate 'substantial funding ... for each local authority to meet non-operational legal responsibilities' (*ibid.*). This funding would have to come from the state. An option for a landfill tax would be included in the Plan's financial recommendations, as would the introduction of private/public partnerships between the local authority and private sectors.

4.4. CONCLUSION: THE STATE'S RESPONSE TO THE WASTE CRISIS: THE CREATION OF WASTE MANAGEMENT POLICIES

The policy agenda of the state has different tiers. In the first instance, the government must operate within the constraints of a policy style that has been shaped by the demands of the EU, corporatist partners, coalition partnerships, and clientelist demands of local government and the wider demand of the electorate who ultimately hold the power of the ballot box over politicians. Another group that were trying to exploit this shift in policy are the multinational incineration industry. This chapter has charted the government's response to these wider pressures, through an exploration of their legislative approach to the waste issue. By attempting to take on all the variables that surround the creation of policy on the waste issue, the state opened up access to political opportunities, which in turn was exploited by GSE.

In this chapter, I detail the state's establishment of the EPA and its eco-modern waste management policy. This waste management framework occurred in the years between the establishment of the EU's own eco-modern waste legislation and the emergence of the Celtic Tiger economy in the decade of the 1990s. To chart the evolution of the state's environmental protection and waste management policies, this chapter has outlined the key junctures in the state's legislative framework.

The first response of the state to both EU legislation and grassroots campaigning was the establishment of the EPA in 1992. The 'top-down' pressure from the EU came in the form of pollution control directives, while the 'bottom-up' pressure from environmental groups during the first phase of campaigning. However, the state's reliance on technocratic advice, combined with an antipathy towards the community campaigns, resulted in an approach to the implementation of pollution controls and waste management policies which could be portrayed as undemocratic and pro industry because of the state's inability to interact with communities with grievances on issues such as health risks of pollution levels. In this way, the state's approach to pollution control and waste management can be seen as being structured to fulfil the requirements or obligations of the 'top-down' directives of the EU, while simultaneously falling short of these requirements during the implementation phase, as part of a technocratic resistance to the 'bottom-down' pressure of community campaigns.

The pollution control and waste management framework established by the state had its inception in the formation of the EPA followed by the policy initiatives such as the *Waste Management Act of 1996* and the EPA *Millennium Report*. By 1999, the state introduced its required waste plans, including the CWP. While the CWP promoted two eco-modern waste management options in recycling and incineration, GSE will be shown to have exploited the opportunity which emerged from the CWP's omission of any mention of the health risks posed by incineration. In that context, we can see the limitations of the eco-modern approach that equates pollution control with a concern with profitability in the waste management process.

This perspective, when combined with the secrecy inherent in technocratic approaches to policy implementation provided the basis for GSE's democratic deficit frame. While the EPA and policy documents represent milestones in the state's policy framework, the contradictions of this policy agenda, which was as much about maintaining economic infrastructure and growth as about pollution and waste management, can be seen. The tension surrounding the state's waste management agenda, which emerged as a response to EU directives and community campaigns rather than from a concern with the environment, was complicated by the state's reliance on landfill. In this context the state set about to create a waste management policy that stayed within the pro-industry remit set out by the EPA, while promoting incineration as a multinational eco-modern alternative to landfill, rather than promoting recycling as a possible eco-modern approach that would appeal to local communities.

NOTE

1. See also: Postscript 2: 'Ecological Modernisation and Irish Environmental Policy' at the end of this book for more details on environmental policy styles.

CHAPTER 5

FRAMING PERSPECTIVES ON WASTE MANAGEMENT: POLITICAL OPPORTUNITIES AND RESOURCE MOBILISATION IN GSE'S CASE

5.1. INTRODUCTION

The first phase of Irish environmental campaigns had exploited the NIMBYist concerns of local communities in a manner that superseded the economic rewards of toxic multinationals promoted by the state. GSE's campaign extended that NIMBY frame by networking with other community groups concerned about the state's approach to waste management in the second phase of Irish environmental campaigning, which was concerned with the post-boom waste crisis and infrastructural sitings.

In this chapter, I explore the emergence of GSE's campaign against incineration, which was included as an option in the state's regional waste plans. In particular, the focus will be on the manner in which GSE mobilised their campaign by framing the issue of grievance such as the potential health risks posed by incineration. The chapter, in particular, highlights GSE's own concern about framing their campaigns in a way that would allow them to avoid being labelled as NIMBYist, while forging links with anti-landfill groups. This can be seen as a type of 'ideological development' (Szasz, 1994) that is also underscored by GSE's attempts to forge wider links with local politicians, through their public debates and meetings. The wider development of GSE's campaign can also be detected through their utilisation of communication technology to create networks. This chapter details these significant events in the emergence of GSE's campaign that saw their campaign emerge from their NIMBY and health risks frame to challenge the state's technocratic approach to waste management.

The campaigns against multinational incineration in Cork and Derry had established a mistrust of the technocratic science employed by the multinationals and the state. GSE were able to utilise this mistrust of technocratic science as part of their frames of health risks and NIMBYism. GSE were aware of the state's responsibility to address the waste crisis through the EU's eco-modern waste hierarchy.

Examining the mobilisation of a campaign's resources, Tarrow makes the distinction between the internal mobilisation of resources such as 'money, power or organisation, and the incentives' that occur within the political environment. This creates the 'dimensions of opportunity' (Tarrow, 1994, p. 83) that the GSE case centred around, such as political and professional expertise. The chapter focuses on this mobilisation of internal resources and also explore the exploitation of the external resources of the political opportunity structure that surround the waste management dispute.

5.1.1. Resource Mobilisation in GSE's Case: Utilising Internal Resources

Studies of collective behaviour have indicated that movement activity occurs after periods of accelerated societal change. Existing political hierarchies, which fail to maintain social cohesion in times of transformation can face challenges by groups that shared concerns and grievances about ensuring crisis. These shared beliefs, when acted upon, are developed into episodes of collective behaviour. In GSE's case, the issue of including incineration as a waste management option created the shared grievance that became the motivation for the collective behaviour of the relevant contenders, comprised of GSE's committee members. The creation of waste policy and the response to GSE's campaign and local authority decisions shaped the governmental approach to the issue. Therefore, we can establish a framework for understanding the POS in this case, with the EU's application of an environmental directive on waste creating pressure from above for the state in regard to its waste policy, while GSE's grassroots campaigns created pressure from below.

The shared interests of its initial committee stemmed from prior environmental and political activism and the creation of public concern about the health risks posed by incineration. GSE's leadership and politically experienced supporters are explored in detail in the next section.

In GSE's case, framing opportunities provided by the experience of this leadership in the utilisation of expertise to underline the potential health risks, and the highlighting of concerns about democratic deficit were all factors in increasing the mobilisation of consensus.

An understanding of the mobilisation of resources and political opportunity structures can be drawn from the GSE case. Here, power was derived from the opportunities afforded through the mobilisation of GSE's campaign. Political opportunities emerged, but then diminished as a result of GSE's involvement in the 2002 electoral campaign. The initial political opportunity arose from the introduction of the CWP, while the mobilisation of internal resources was facilitated by the activities of personnel with prior experience and expertise, as interpersonal contacts were activated between GSE's committee and US anti-toxics and Irish anti-incinerator and 'superdump' groups. Further political opportunities were created through the state's ignoring the results of GSE's petition, and the councils' rejection of incineration. These events created a wider political opportunity structure for GSE to utilise, with certain events giving the group an advantage or disadvantage in relation to their challenge to government policy.

GSE's organisation grew from GSE's personnel, through their shared interest of political and environmental activism. Mobilisation occurred around the utilisation of resources, such as prior experience and political and technical expertise, to publicise the group's cause. This expertise was used to highlight the health risks posed by incineration for local communities. Further opportunities were provided from the networks created by the group. GSE mobilised the internal resources available to them by building a campaign of public events and media debates to promote their main frames.

Taking a closer look at GSE's committee, the peripheral nature of the participant's politics (i.e., splinter group republicans, or anti-incinerator activists in an incineration-free country), it could be said that they were comprised of a potential leadership without a cause, until the issue of incineration emerged as a political opportunity to be exploited. The group could also be accused of having a localised, NIMBYist perspective. GSE would address these concerns about NIMBYism as part of their framing process. There were a number of environmental or left wing activists with prior experience in the Galway area, having participated in disputes concerning sewage treatment plants and infrastructural conflicts in the nearby County Clare. In addition, the existence of two third level institutions in the city of Galway is a significant source of educated people with a degree of understanding about environmental issues.

GSE's initial meetings were held in the Castlegar Community Centre, near one of the named sites for the proposed incinerator. These meetings were organised by Conchúr Ó Brádaigh, an engineer and lecturer and political activist who lived nearby. The first guest speaker was Aine Suttle, who had contacted concerned residents in the area to offer her expertise, gained in Canada, where she had been an activist in an anti-incineration movement. In GSE's case, the legitimation of action goals was achieved quickly, as experienced political activists began to shape the direction of the campaign. An initial attempt to have the campaign taken over by the local Green Party was resisted, as the committee wanted to avoid association with any one party, while the Greens were not a strong group in Galway at that point. This resulted in the resignation of the Green party representative from the committee. Good relations with local and national media groups were established, through the provision of accurate material for journalists and editors. Wider networks were established, with national and international figures. A strategic plan was drawn up, with achievable goals, supported by meetings. Communication technologies were utilised by professionals who had the expertise necessary to exploit the information flows of the Internet. In this way, consensus was mobilised around GSE's ability to present a credible and professional campaign.

This section has looked at the mobilisation of GSE's campaign, by examining the political opportunity structures, and the mobilisation of resources and consensus that were undertaken by the group. The chapter will go on to examine the leadership elite that came together to direct the campaign, both internally and externally. Together these people were able to influence responses to the issue of incineration, to an extent where policy could be challenged through an interaction with the political process.

5.1.2. Framing the Issues: Concerns about NIMBYism

The announcement of the CWP's option for incineration in named sites such as Castlegar resulted in an immediate response from the community. A large number of the GSE committee were from the Castlegar and North Eastern section of the city, leaving GSE open to the accusation that they were NIMBYs (Not in My Backyard). Ó Brádaigh realised the need to avoid the 'NIMBY, label:

> We made no reference to site selection; we were just opposed in principle to incineration. Although we use the Castlegar area as our base, we consciously don't have meetings in the Castlegar Community Centre. We don't want to be as a sectional group ... people get

> involved in what affects them directly. It's democracy. I mean, you can be accused of being a NIMBY, but, I prefer to say 'Not On My Planet' (ibid).

Suttle had experience of the accusation of NIMBY self-interest being directed at anti-incineration activists in Canada and urged GSE to try to avoid being depicted in this manner. NIMBYism is analysed as a feature of waste protest by Szasz. He states:

> The hazardous waste and toxic movement is often characterised by its critics in industry and government as a NIMBY, or 'not in my backyard' phenomenon.

Prior experience was made available to GSE through the expertise of anti-incineration campaigners such as Aine Suttle and Paul Connett. It is the availability and utilisation of this experience and analysis that shaped GSE's campaign. The transition of GSE from a NIMBY-style campaign of concerned citizens in the Castlegar area to that of a city and county-wide campaign which developed an analysis of the waste crisis began in earnest with a series of public meetings scheduled for the first three months of 2000. These meetings were held in the Corrib Great Southern Hotel, and the Menlo Park Hotel, located near the Castlegar area. According to GSE, this good relationships with both the local and national media was based on a determination to steer GSE around the perceived pitfalls of other recent environmental campaigns in Galway, in particular, the 'Save Galway Bay' anti-sewage treatment protests that had occurred between 1997 and 1999.

That campaign had been widely perceived as fundamentally NIMBY in its character. Residents of the Claddagh and Renmore areas of Galway had clashed over the siting of a proposed sewage treatment plant, which was to be sited in Galway Bay at two potential sites near either neighbourhood. The subsequent campaign was protracted and heated, and local press editorials were strongly against the protestors, who were portrayed as being narrow minded and self-interested. Ultimately the case was heard at the High Court, with Justice Keane finding against one of the leaders of the 'Save Galway Bay' campaign Ray McBride, costs to the amount of millions of pounds.

When interviewed Ó Brádaigh indicated a related series of factors which helped GSE in the transition from what may have been perceived as a local-based NIMBY group to that of a more widespread environmental protest movement. This was due to the 'fundamental' nature of 'the issues involved: health, children's health, economics, cost, recycling' (C. Ó Brádaigh, interview, May 2000). GSE drew support from all sections of society.

> I'd say that the structures that make GSE effective are the fact that its heavily based in the community, totally voluntary effort. You have a mix of people from all sorts of backgrounds, all parts of the city, rural/urban, politically involved and not politically involved. (*ibid.*)

5.1.3. Mobilisation of the Internal Resource of Leadership

The leadership of movements and groups is a critical aspect of the mobilisation and strategic performance of their campaigns. Leaders make decisions that shape the direction of movement challenges. Leadership helps to provide a degree of competence for movement organisations. In GSE's case the existence of personnel with an array of leadership qualities and relevant expertise became a key resource. This section examines the participants in GSE's wider leadership group, those who worked within the campaign or provided external support.

In a conscious attempt to avoid the difficulties encountered by the 'Save Galway Bay' leadership, GSE didn't appoint a named spokesperson. Instead, they utilised the informal approach of new social movements (Martell, 1994, p. 112), which are based on the mobilisation of expertise networks. However, it soon became clear that Ó Brádaigh and Suttle were the most vocal committee members. Leaders play an important role in 'the accumulation and allocation of resources' (Klandermans, 1989, p. 216). They can provide the momentum for the mobilisation of consensus required to launch a campaign. GSE's committee assumed this leadership role for the duration of their campaign.

However, it could be said that GSE's lack of a clear leader may have led to problems for their campaign, as the case study will demonstrate. Over time, Suttle and Connett would come to differ on GSE's third frame of highlighting concern about democratic deficit during the 2002 General Election campaign. However, at this early stage of GSE's campaign, it was the prior experience of its committee that proved to be a significant resource, which facilitated the growth of the campaign. By providing a credible basis for their challenge, this group of professionals steered the campaign's emergence into the political arena. The primary focus for the campaign was centred on a challenge to the state's policy on incineration with further opportunity for expansion located by the leadership in the CWP's omission of potential health risks, in addition to the lack of wider public consultation on the issue. The prior political experience of some of the leadership group, in addition to the networks it established, gave GSE the basis for an effective campaign against the state's waste management policy.

GSE's leadership shared a belief that its campaign could result in a successful challenge to the state's waste policy. Beyond the shared grievance of the imposition of an incinerator, the leadership group also held a belief that it had the relevant expertise necessary to challenge the state's waste management policy, and offer a credible alternative in the shape of a more extensive use of recycling. The concern expressed by groups or members of the public on environmental issues tends to emerge from pollution risks in their locality. As such, forms of NIMBY or 'not in my backyard' resistance at local level can stay as just that, or subsequently form links with existing environmental groups, as part of a wider campaign against specific environmental degradation. Media interest underpins public attention to environmental issues, granting a degree of perceived social importance that can go beyond an environmental issue's immediate local relevance. As an environmental group's concerns become highlighted through media debates and public campaigns, other environmentalists are able to provide assistance and expertise based on previous campaign experience.

GSE identified the following people as suitable committee members. Aine Suttle had the largest contribution to make to the group, due to her personal experience and commitment to campaigning against incineration. Her links to a global network of anti-incinerator groups was to be the most valuable resource of the group. Ó Brádaigh himself could provide the political expertise to lead the campaign, and he had media experience from his work with Republican Sinn Fein in previous campaigns. Other RSF activists would provide organisational skill to the group. RSF members were part of many local societies in Galway, and even their activities on the church committee in his local parish would provide further alliances with the community. They also had the expertise to establish a website for GSE. Additional support was provided by some prominent businessmen living in the area, and these figures could provide links with the business community. Other local community activists would provide their knowledge of local community politics, through his long-standing expertise as a resident's association leader. The collective application of this group's political, technical and professional expertise would provide the main resource of the campaign. The group adopted the name 'Galway for a Safe Environment', and the committee members began to plan a campaign that would utilise Castlegar's position as a rural community on the edge of Galway City to link both the rural based anti-super dump campaigns based in Newbridge, New Inn and Kilrickle in the county, with concerned citizens and environmental groups in the city. As Suttle put it:

> An incredible group of people got together with the most perfect merge of skills. I had the information, Conchúr Ó Brádaigh had the media savvy ... Joe Jacobsen took on fundraising; he has those business skills and contacts. Paul O'Malley became joint secretary with me. There were people there who were willing to fill the gaps. We all seemed to fall into roles very naturally. (Aine Suttle, interview, March 2002)

The newly formed committee of GSE also made a number of decisions that would facilitate the mobilisation of a campaign against incineration. The committee decided to meet every fortnight, to plan and implement their campaign. Different committee members were assigned specific tasks, in relation to the expertise they had at their disposal. The relevant expertise of the committee would be utilised to create the frames of health risks and NIMBYism. Ó Brádaigh would deal with the media. Suttle began sourcing material about the potential health risks of incineration, which would be presented in a pamphlet, 'A Burning Issue'. Some committee members would lobby local community groups about a meeting on the issue. Others were put in charge of fundraising and websites. Ó Brádaigh would use his political contacts to begin a campaign to push the deadline for submissions back. This deadline was originally set for March 2000, but the group felt it would be important to increase levels of participation, empathy and publicity for the group.

The main link with existing environmental groups for GSE was Aine Suttle. As a volunteer at the Galway One World Centre, Suttle had tried to create links with environmental groups through the Internet, but had found that process relatively limited. Originally from County Mayo, Suttle had lived in Canada for twenty years. In Toronto, she had become involved in anti-incinerator campaigns. This had provided her with a wealth of scientific data concerning the health risks of incineration, as well as a good deal of literature and knowledge of strategies that GSE could utilise. Perhaps most crucially, Suttle provided the link with Dr. Paul Connett, who has been described as the leading world expert on dioxins; the main toxic pollutant associated with incineration (Walsh et al., 1997). While individual contacts were established, she found it difficult to locate any active environmental groups around Ireland. Crucially for GSE's campaign, Suttle did make contact with Joe Jacobsen in Castlegar, which led to her becoming the first speaker at GSE's inaugural meeting in December 1999. It also brought her into contact with Joe Bridges of the Wexford anti-incinerator campaign. Bridges was later to speak at a GSE public meeting, and the Galway-Wexford link was the beginning of an overall coalition of anti-incinerator and anti-super dump groups around Ireland.

While Suttle found the processes of both the Irish and Canadian anti-incinerator campaigns similar, she detected a major difference between the reactions of politicians in both countries. Suttle's first campaign was in Toronto. In that city, a petition of 6,000 signatures from a city of two and a quarter million was enough to convince local politicians of the strength of opinion that had formed against incineration. According to Suttle, this petition had "a big impact", despite the small proportion of signatures per head of population, as the Toronto authorities dropped their plans for an incinerator after the campaign. On the strength of this outcome in the Canadian case, Suttle suggested a similar petition for GSE's campaign. This petition amassed 22,000 signatures from a population of 65,000. Despite this large response, local politicians dismissed the GSE petition. For Suttle this demonstrated 'that politicians here are, far less responsive to their constituents'. Another difference in the political process of the two countries noted by Suttle was that when politicians in Canada were lobbied 'they tended to be frank with us', whereas Irish politicians are less forthcoming about where they stood on certain issues. "I don't understand this, I'm Irish, but they make you jump through hoops that are utterly useless". Suttle also indicated that in Toronto, the Local Authority funded protest groups, welcoming their contribution. In Galway, Suttle found city councillors to be in the main dismissive of GSE. This indicates two differing official responses to environmental protest, one (Toronto) addressing the democratic deficit, and the other (Galway) adding to a sense of public alienation from the political process.

5.1.4. Health Risks

The first of GSE's frames involved the highlighting of the health risks of incineration. GSE embarked on a series of tactics to promote this frame, through a campaign of public events and media debates. GSE used tactics such as public meetings, drafting a pamphlet, utilising Internet communications to facilitate news media interaction and networking with other groups and individuals opposed to incineration. The primary focus of these tactics was the highlighting of the health risks posed by the dioxins and furans found in incineration emissions.

One of the early tactics undertaken by GSE was the publication of a pamphlet providing GSE's arguments against incineration, backed up by scientific information provided by Suttle from Dr. Paul Connett, the US anti-incineration campaigner. The publication of a pamphlet was Suttle's idea and stemmed from her experience of this tactic during previous anti-incineration campaigns in Canada. This pamphlet was titled *The Connaught Waste Management Plan, "A Burning Issue"* and highlights the broad parameters of discursive analysis that existed before GSE's campaign, particularly around the area of the health risks posed by the emission of dioxins from incineration. The use of scientific information throughout *A Burning Issue* indicates GSE's ability to frame the expertise available to the group from different campaigns in Ireland, Europe and North America. This form of scientific expertise has been described as 'interest driven' by Grove-White (1993, pp. 22–23). Such scientific data form the bases of interest-led challenges to the technocratic data provided by experts for multinationals and the state. The framing of health risks posed by dioxins became an important part of GSE's anti-incineration message, and was highlighted throughout this stage of the campaign. By exploring GSE's pamphlet *A Burning Issue*, an understanding of the importance of the expertise and links to previous campaigns of GSE's committee members can be provided. What becomes clear from this examination is the crucial nature of GSE's ability to link with Dr. Paul Connett. This link provided GSE with the necessary interest-led scientific expertise to galvanise their campaign. GSE's pamphlet highlighted the broad parameters of discursive analysis that existed before GSE's campaign, particularly around the area of the health risks posed by the emission of dioxins from incineration. The pamphlet begins with a summary of *The Connaught Waste Plan*, which was published in September 1999. Three areas of waste disposal are set out in the CWP. These were:

> the diversion of 50% of overall household waste away from landfill, a minimum of 65% reduction in biodegradable wastes consigned to landfill, and the recycling of 35% of municipal waste. (*A Burning Issue*, 1999, p. 1)

The third recommended option for waste disposal in the *Connaught Plan* provided GSE with their most serious issue of contention. It states that a third option would include an

> Incinerator located near Galway City, to burn approximately 184,000 tonnes of rubbish per annum, by 2013 (44% of waste by 2013 without construction waste. Four possible sites identified, Castlegar, Brockagh, Coolagh/Ardain and Galway docks. Recycling initiatives to handle approx. 160,000 tonnes per annum by 2013 (38% of waste), residual landfills handling approx. 80,000 tonnes per annum (19% of waste) possible sites

> identified as Newbridge, New Inn and Ballinahistle (Kilrickle), total investment cost in the region of £300 million. (*ibid.*)

The identification of possible sites for an incineration plant and residual landfill alerted local communities in these areas and led directly to both the formation of GSE in the Castlegar area, and anti-superdump groups in Newbridge and New Inn. Despite empirically opposing interests, these groups would link successfully to form an overall protest against both aspects of contention contained in the waste plan, incineration and landfill.

By identifying the potential sites for the incineration plants, the CWP had provided GSE with the opportunity to 'increase access' (Tarrow, 1994, p. 86) to political structures. The next section of *A Burning Issue* contains GSE's 'General Criticisms' of the waste plan. Here the main themes of GSE's campaign can be identified. Of particular concern is the lack of any mention of 'health issues', a point that GSE was to use against the authors of the waste plan throughout their campaign, as part of their health risks and democratic deficit frames. This point is reinforced by the highlighting of the lack of any reference to the threat posed by dioxins, the harmful chemicals contained in incinerator emissions. The word 'DIOXINS' was presented in bold print, and became one of the main buzzwords of GSE's campaigns. The pamphlet highlights the health risks of the toxic ash residue created by incineration. The waste plan is also faulted for its 'sketchy details on economic forecasts' (*ibid.*).

Further economic criticism of the waste plan was provided through highlighting the relative costs of recycling as opposed to incineration.

> Even though recycling is cheapest, it was written off as not being able to meet the national targets, despite the fact that many places in the world are surpassing those targets. (*ibid.*)

While the 'General Criticisms' section of *A Burning Issue* provides an indication of the importance of Aine Suttle and Paul Connett in providing the scientific information concerning the health risks posed by incineration, the next section of the pamphlet titled 'Our Conclusions' highlight the local concerns of residence in the Castlegar and Newbridge areas. By establishing a rapport with residents in both these areas, Conchúr Ó Brádaigh came into his own as a strong leader for the groups, despite his reluctance to be perceived as an overall spokesperson. Through his extensive political and academic links, GSE were able to form an alliance with the rural-based anti-'super dump' protestors. This was a deliberate ploy to prevent the Local Authorities from driving a wedge between both groups, by using the rural urban divide that exists in Irish society for their own means. In this way, GSE began to expand beyond their NIMBY frame.

GSE were aware that the tactic of dividing communities had been successfully introduced during the Galway Bay/Mutton Island Sewage Treatment Plant campaign. At that time, the local residents of the Claddagh and Renmore areas of Galway City were pitted against each other in a debate over the siting of a treatment plant. These divisions created an embittered and divisive campaign, one that deflected the attention of the public off the environmental issues at hand, and away from the perception of the Local Authority as the main protagonist of the issue. The pamphlet's emphasis then switched from health risks to items of a more political nature. The first item accuses the state of being undemocratic when it states

> The Connacht Draft Waste Management Plan is an attempt to stampede the people of the West into accepting an incinerator in or near Galway city. (*ibid.*)

By using the phrase 'stampede', GSE attempted to portray local and national government as being undemocratic on the issue of Waste Management. This becomes the basis of a charge of lack of public consultation, selective discussions about the health risks involved, and the need for a broader debate on all areas of waste management and recycling. This part of GSE's framing of the democratic deficit issue represents their initial attempts to expand their campaign beyond NIMBYism into a political challenge to the state's policy framework.

5.1.5. Framing the Issue: The Utilisation of Communication Technology

This section explores the framing of issues by movements, by exploring the impact of such links between GSE and anti-incineration experts who have provided vital information and strategic material, allowing issues to be framed in a particular manner. Castells defines Internet technologies as a 'privileged tool for acting, informing, recruiting, organising and counter dominating' (2001, p. 137). Through the application of communications technologies, localised environmental movements can link with similar groups on a global basis, learning from the hard-earned experiences of previous campaigns. This has transformed the once isolated pockets of environmental resistance into a global movement able to challenge transnational corporate polluters, in the place of regional campaigns of resistance. The increasing reliance on communication technologies by environmental groups has seen a transformation in activists' profiles; changing from the politically dogmatic campaigner to that of a media-friendly professional with a high level of expertise in a related field. This use of expertise has created what Castells has called 'the new dynamics of social movements' (2001, p. 138).

The Internet has become the activist's meeting house in the information era. Networks of opposition have been established and continue to reinforce protests globally, with such success that industries and institutions are now attempting to restructure their own practices in response to the new cultural and political expression of environmental values. This process of redefining cultural values has traditionally been part of old social movement agendas; however, Internet technologies make this redefinition or reinterpretation of values much easier for relevant movements today.

This shift in social movement capabilities can be seen in the transition of protest groups from a reactive force for change within a dominant ideological paradigm; (i.e., civil rights granted by the perceived institutioned oppressor) to that of a cultural movement which can set the agendas and frameworks of their campaign (Greenpeace's influence of public perceptions of the environment). In turn, old political conflicts can have their overall boundaries redrawn in this era of communication technologies, as cultural values are redefined through information flows. In this way, old political values (and conflicts) can be resurrected through a campaign based on communication of new perspectives.

The restructuring of political and cultural values through the technologies of communication has become an important strategy for social movements in their attempts to create collective action frames. It is what Castells refers to as 'mobilisation around meaning' (2001, p. 140). The importance of new political understandings of meanings becomes evident in relation to the crisis of legitimation facing democratic politics. As mainstream politics veers towards a centrist, Liberal Democratic monolith, radical political expression had adopted new organisational forms, with traditional hierarchies replaced by Internet links that can give equal voice to a multitude of concerned activists, each able to give as much or as little to a campaign as Internet technology allows.

Furthermore, movement campaigns are now being structured around the available technologies of communication. The medium can affect the message, and often shape it. For instance, a movement's success or failure may be based on media depictions of certain aspects of a campaign, rather than being perceived on any subsequent outcome. Issues are framed by media images, such as Greenpeace running a flotilla in the Irish Sea, alongside ships carrying a nuclear cargo for British Nuclear Fuels (BNF). Resolution to the vexed question of nuclear power isn't achieved or expected from this event. However, Greenpeace gains a large amount of public support for their movement from these images. Ultimately, the media is utilised to create transferable emotiveness around a political event.

For protest movements, the Internet allows for a strategic diversity which prevents such movements from becoming stifled by drawn out campaigns which can sap morale. Through communication technologies, the battleground of movement protests can be shifted at the press of a keyboard button. This gives protest movements the ability to link and mobilise with allies in cities around the world. It also eliminates any chance of sustaining a perceived defeat on an issue, as even the most heavy-handed response from the authorities while inflicting physical harm to front-line protestors, provides the type of publicity necessary for movements' success in the long term.

The ability to frame issues by movements has become an important aspect of the politics of protest. Strategically, the mobilisation, coordination and publicity needed for protest movements at a global level can only be achieved through communication networks. What becomes clear from an analysis of how protest movements utilise communication technologies is the increasing importance of these new forms of technology as a tool for framing issues and organising campaigns. Technologies of communication ultimately create a platform for movements that can frame issues around shared grievances.

Essentially, the strategic impact new media forms have had on protest movements has been based on a movement's ability to frame issues that influence overall media agendas. The ratings potential of protest coverage is dependent on two main themes; framing the message of the movement, and the potential for an attention grabbing media event emanating from that movement's activities. Therefore, the manner in which an issue is framed becomes a critical aspect of how media attention is maintained. According to Rogers and Dearing (1988), there are three main features in the issue maximisation process. These are part of the framing process that incorporates communications elites such as multimedia corporations, movement events that embrace the spectacular, and a coordination of media, public and policy agendas.

From an Irish perspective, the utilisation of media and Internet has become a central feature of the framing process undertaken by GSE. One of the most interesting aspects of GSE's campaign was their utilisation of Internet technology and expertise to frame their campaign around the potential health risks posed by incineration. By utilising information, such as the 2000 US EPA report on the dangers of dioxins from incineration,

GSE were able to transfer information through what Castells has called 'global computer-mediated communication (CMC)', in order to frame the 'moral discourse' (Grove-White, 1993, p. 22) surrounding the incineration debate. A GSE committee member, discussed this point in the course of an interview.

> The internet has had a big impact on this and other environmental campaigns. A good example of this is that on the day the US EPA report was released, we were able to have it from the front page of the Washington Times that morning ... hearing about this at one o'clock (Irish Time), to getting Dr. Paul Connett on the Six One news, through internet connection ... Its remarkable that the consultants involved in the Galway Waste Plan have a highly paid P.R. firm working for them, and yet, I'd say without beating our own drum, that we've beaten them hands down in the local media. (C. Ó Brádaigh, interview, May 2000)

GSE were initially successful in their utilisation of both local and national media in regard to framing the health risks debate, as committee members maintained that a run of well received television appearances, particularly one where they debated the incinerator issue with former Taoiseach, Albert Reynolds, were successful, due to their ability to present expert information, which allowed journalists to trust their perspective on the issue. This strategy was based on 'contacts and other campaigns', where GSE picked up on the maxim 'source everything you use. If you can't source it, don't say it' (*ibid.*).

What surprised GSE, and aided their framing of the health risks issue, was that 'the consultants didn't adopt the same standards, and tended to throw sweeping statements' (*ibid.*). From this, the local print media began to show editorial support for GSE's campaign, despite a history of opposition to protesters in previous campaigns, such as the divisive anti-sewage-treatment plant campaign in Galway during the early 1990s. By challenging one form of technocratic or 'official science' with their 'interest driven science' (Grove-White, 1993, p. 21), GSE's movement entrepreneurs were able to frame the health-risk issues in a way that achieved a degree of consensus amongst local politicians and concerned members of the public. By exploiting the public's concern about potential health risks, GSE had framed the issue in a way that allowed for further campaign mobilisation.

> The day has gone when highly paid consultants can come into a community and tell them ... that this is perfectly safe. I mean, it took one of our neighbours, a librarian, to find in five minutes on the internet, a great number of highly damaging facts in recent scientific publications. These articles gave a lie to what the consultants were telling us. (C. Ó Brádaigh, interview).

GSE's use of the US EPA's Report on the health risks of dioxins represents another successful framing tactic, as international links were used to frame the dioxins issue in a wider context. What GSE's campaign has demonstrated is the importance of professional experts who can outflank both government authorities and private consultants, leaving environmental disputes to be framed and contested in the field of media primetime and Internet cyberspace.

5.1.6. Mobilisation of Networks

GSE were able to mobilise the internal 'interpersonal links' (Friedman & McAdam, 1992) of their members. GSE were intent on linking with the resource of rural anti-superdump groups, to frame the waste issue in a way that combined the societal grievances of both rural and urban citizens. GSE argued that recycling was the best answer to the waste management crisis in an attempt to frame the waste issues as one of being recycling versus incineration and landfill, which would maximise their public appeal. GSE presented recycling as the best option throughout the debate, and highlighted the fact that recycling removes the vast amounts of waste needed to fuel an incinerator from the process.

The value of Suttle's Canadian experience on this framing of the waste issue for both the anti-landfill and anti-incineration groups was recognised by other GSE members:

> Aine Suttle ... has experience of campaigning against incinerators in Canada ... and her experience was that incinerator companies had actually driven a wedge between the people concerned about air pollution and those concerned about ground pollution. They actually funded the anti-landfill groups, and this created a lot of trouble for environmental groups. (C. Ó Brádaigh, interview, May 2000)

GSE's aim was to reach out and construct links with all the disparate environmental groups in the Galway regions, who had previously been working in isolation from one another, and as such, had gained little success in their campaigns. GSE took the initiative in establishing networks with the anti-superdump groups, as well as with community groups in the city. These groups were willing to form an alliance with GSE that strengthened their cause in the city and county.

Further links were established with Dr. Joe Bridges of the Wexford anti-incinerator campaign. GSE used information from Dr. Bridges to frame the issue around the threat to agriculture that incineration posed. Bridges and

Suttle had been in contact online in the months before GSE's campaign commenced. They exchanged information on their campaigns, and Bridges was later to speak at GSE's public meeting in the Menlo Park Hotel.

The existence of expertise networks strengthened GSE's framing of the issue. These links, with national and international anti-toxics groups, allowed GSE to rise above accusations of NIMBYism and provided an effective platform for the framing process which would sustain the initial phase of GSE's campaign. Essentially, the effective utilisation of expertise by GSE's committee allowed it to mobilise the discontent of a public concerned about environmental issues and angered by political incompetence and corruption. By expanding their NIMBY frame in this manner, GSE were able to eventually move their campaign in a transition from NIMBYism to a concern about democratic deficit, this became their third frame.

5.1.7. Framing the Health Risks of Incineration

One of the ways that GSE tried to frame the issue of health risks was through the publication of a pamphlet titled '*A Burning Issue*' in December 1999. *A Burning Issue* provides detail on the problems of incineration. In all, 10 headlines are provided, each of which discusses aspects of GSE's opposition to incineration. These points represent GSE's framing of the issue, around what would become the key themes of GSE's campaign The first of these frames can be seen in the statement claiming, 'Incineration is a health risk' (*A Burning Issue*, 1999, p. 1). Burning rubbish, it is claimed produces 70% to 80% of emissions in the shape of 'gases, vapours and fine particles' (*ibid.*) Similarly incinerators emit dangerous chemicals, including dioxins, furans, hydrogen chloride, nitric acid ... particles of ... lead, cadmium, arsenic, mercury, chromium, etc. (*ibid.*). By presenting details of scientific data about the health risks of dioxins, GSE began to frame the issue as a contestation between issue led science and the technocratically led 'official science' of the state that could be portrayed as untrustworthy, as the state had not dealt with health risks in the CWP.

5.1.8. Framing Understandings about Dealing with Waste

The framing of health risks was extended into a wider debate about waste management processes. Waste management issues are introduced in the pamphlet under the heading 'Incineration does not eliminate the need for landfill' (*ibid.*). As incinerators produce a fly ash residue that must be landfilled, GSE extended their framing to again link the issues of incineration and landfill. The statistic of 1 tonne of ash being produced by burning 3 tonnes of rubbish is presented as evidence of the potential problem fly ash poses, one which provides 'an even greater hazard to land and water than the original landfill' (*ibid.*). The siting and longevity of such landfills is also raised, as the framing process came to underscore GSE's tactic of linking their anti-incineration stance with concerns about landfill.

GSE also framed their argument around costs, claiming that recycling is more cost-efficient than waste incinerators. To support this, the *Connacht Waste Plan* is quoted as stating that 'the recycling option is 25 million pounds cheaper than other waste management combinations which included incineration' (CWP, 1999). The efficiency of incineration is also challenged. German studies are quoted to demonstrate that incineration 'contributes only half of a percent of the countries energy requirements' (*ibid.*). A crucial point made concerning 'tipping fees', or the payment made by local authorities to incineration plants. While 10% of an incineration plant's income is from energy generation, 90% comes from such 'tipping fees'. These fees are stated to be as high as $100 to $200 per tonne in the United States and Germany, according to Dr. Connett's 'Waste to Energy' Report. GSE highlighted their economic arguments, by framing the costs issue in a negative light:

> Incinerators are the only form of power-plant which gets paid to accept the fuel it burns. (*A Burning Issue*, 1999, p. 4)

In addition, GSE argued that the numbers employed in incineration plants are very low. According to the *Galway Waste Management Strategy Study*, Galway's proposed incineration plant would employ only between 15 and 50 people, a small number compared with the high employment created by recycling sorting plants. The profit base of the incineration industry is another area criticised by GSE. They argued that multinational incineration industries would remove their profits from Ireland, while leaving their unwanted pollution behind. GSE argued that recycling tends to be localised and community based and that profits remain in the area where the recycling takes place. In a final economic critique, GSE stated that the consultant for *Galway's Waste Management Strategic Study* claimed that the maintenance on incinerator plants is very expensive, and that 'incinerators need to be retrofitted every ten years'. According to GSE, this raised the spectre of an ever- increasing number of incinerator plants being required to replace defunct or ageing plants over a number of years. In this way, GSE framed the waste issue around economic factors, arguing that incineration was expensive and inefficient, while recycling creates local profits.

The clear alternative to incineration, according to GSE, was recycling. In a further criticism of incineration, GSE argued that the cheaper and more ecological recycling presents is threatened by its competition with the international incineration industry. They asked the question 'will long term contracts guaranteeing rubbish deliveries have to be signed' (*A Burning* Issue, 1999, p. 4)? In other words, will recycling be actively discouraged in order to provide materials for incineration plants? GSE also point out that *The National Waste Management Plan* echoes the EU's waste hierarchy reiterating the preferred options: reduction, reuse, recycling and composting with landfill and incineration as a last resort.

Canadian statistics are quoted as saying as much as 80% of waste is suitable for compost and recycling, with many North American cities going as high as 50% in their recycling efforts. In Canada, this figure is set even higher at 60%, according to the Nova Scotia Department of the Environment. GSE note that the original preference for the Canadian authorities was incineration, but a concerned campaign of resistance from 1994 led to a change in the direction taken by the Department of Environment.

GSE also framed the issue by using data from abroad, backed by data from other countries' environmental and anti-incineration campaigns: 'Citizens in other countries won't allow incinerators to be built, so why should we? This is followed by the following information from Dr. Paul Connett:

> In the USA in 1990, 140 incinerators were operating but between 1985 and 1998, over 300 incinerator projects were cancelled. The incineration industry in the USA is dying. Some operating incinerators were prematurely shut down, such as the one at Glen Cove, Long Island, which closed in 1991 after only 8 years of service. (Paul Connett, Menlo Park Hotel, January 2000)

The mobilisation of expertise and links by GSE allowed their arguments to be framed and presented in a way that would maximise the support of the public, creating the leverage necessary to gain access to political opportunity structures.

By using the professional expertise of their committee members, GSE established Ireland's first anti-incineration website, with connections to US anti-toxic and environmental movement websites. Print, media, radio and television support were achieved through a professional presentation of GSE's arguments, based on scientific facts. By providing the media in

Ireland with the material for an overall debate on the incineration issue, GSE gave the issue a topicality and allowed the media to present GSE's arguments with an equal weight to their opponents.

One way that GSE framed the waste issue through the media was with live television debates. GSE's initial campaign was notable for the success of these televised debates on the national broadcaster, RTÉ. In one memorable debate live on RTÉ's Main Evening News, GSE spokesperson Conchúr Ó Brádaigh effectively debated the incineration issue with a former Taoiseach, Albert Reynolds. While Ó Brádaigh appeared outside his local national school in Castlegar, well prepared and able to present scientific data on the issue, Reynolds looked uncomfortable answering specific questions he had no answers for. His final response to Ó Brádaigh's questioning, Reynolds said 'that's fine, if you don't want incinerators in Ireland, they'll find somewhere else to build them' (A. Reynolds, *Six-One-News*, 12 February 2000).

GSE's ability to frame the issue in an effective manner led RTÉ to continue to cover the debate with interest throughout the subsequent campaign. Ireland's second television channel, TV3, also covered the anti-incineration debate. GSE's public meetings at the Corrib Great Southern were also the subject of a live outside broadcast on that evening's nine o'clock news. This amount of television coverage, combined with a favourable local press coverage that had been denied to previous environmental campaigns in Galway, led to an increase in public support for the group. Local politicians were now being contacted by large numbers of a concerned public, as the health risks of incineration presented in the media by GSE began to take hold with the local community.

5.2. THE MOBILISATION OF WIDER SUPPORT FOR GSE

GSE had succeeded in framing the health risks through the initial phase of their campaign. Other tactics employed by the group proved to be successful in their mobilisation of new and existing activists in the Galway area. One of GSE's most successful tactics was the gathering of a petition, signed by 22,000 people out of a city of 60,000. However, the petition gathering itself proved to be a successful mobilisation process as large numbers also signed up to join GSE and began to attend their public meetings having signed the petition initially. Street signs, hoardings and advertisements were put up around the city and country, which presented the public with a stark image of a skull with the logo 'dioxins kill', creating a potent image of the dangers of incineration in the public's imagination. Existing political links provided GSE with support from many of Galway's active resident associations, who were formally invited to all GSE's public meetings.

In contrast to GSE's effectiveness at this point of their campaign, state and local government officials seemed to be caught off-guard by the strength of the campaign against incineration. Many local politicians were largely supportive of GSE, as well as some rural councillors. The spokesperson for the consultants for CWP, P. J. Rudden of M. C. O'Sullivan, appeared to be unable to provide a defence of incineration that could meet increasing public concerns, due to his technocratic background. GSE's ability to utilise previous campaign expertise allowed the groups to counter many of the authorities' potential tactics, such as driving a wedge between urban and rural communities on the incineration versus landfilling issue. GSE provided an alternative to this argument, by emphasising the importance of recycling processes. GSE were at this point 'winning the debate hands down' (Ó Brádaigh, interview, 15 May 2000) due to this successful framing of their argument, networks and media utilisation.

5.2.1. Setting the Agenda: Public Meeting January 2000

GSE began to mobilise their campaign in early 2000. The first task for GSE was to have the submission deadline for the waste plan put back. Local resident's groups were contacted and local media was used to publicise meetings. A meeting was arranged with local residents groups, to gain their support. To build on this momentum, a public meeting was scheduled for the Menlo Park Hotel. By January, campaign momentum on the ground, together with some effective media appearances by Ó Brádaigh, provided GSE with the opportunity to hold a large public meeting in the Corrib Great Southern Hotel. The meeting was advertised in the local press under the banner headline: 'Stop the Incinerator'. This was followed by an invitation to an evening with Professor Paul Connett from the US, the foremost expert on incineration and related health risks and advocate of the 'zero-waste strategy' being adopted worldwide. The advertisement included the following:

> Incinerators are being shut down worldwide. Incineration is the major source of dioxin in the environment-also produces deadly mercury, lead and cadmium, does not eliminate the need for landfills. Every three tonnes of rubbish produces 1 tonne of toxic ash-to be landfilled in East Galway, is very expensive-£150 million plus to construct. Refuse

> charges likely to treble- only 15–50 jobs created, is in direct competition with recycling, for paper, plastic and cardboard, £150 million would create up to 1,000 sustainable jobs in recycling. (*Galway Advertiser*, 9 March 2000)

The health risks and costs of incineration were to become the main themes of GSE's campaign and were reiterated at their public meeting held at the Corrib Great Southern on 14 March 2000. GSE were mobilising the internal resource of expertise to propagate their concerns to the wider public.

At the meeting, the audience was introduced to Professor Connett, who had experience with anti-incinerator campaigns globally over the previous 15 years. Connett highlighted the fact that the Connacht Draft Plan had omitted my mention of the health risks posed by dioxins. He found this 'incredible given the large amounts of research data available worldwide' (P. Connett, interview, 23 January 2000).

Also addressing the meeting was Dr. Ó Brádaigh, who called for an alternative waste plan, based on 'a partnership approach between elected officials and the community'. GSE called for greater discussion on alternative waste management methods such as recycling and waste separation and likened the anti-incinerator groups, which were largely urban based, with the rural anti-superdump movement. He also challenged P. J. Rudden, (the spokesperson for M. C. O'Sullivan, the waste plan consultants) to an open debate with GSE on the waste plan, to be chaired by an independent figure. GSE also claimed that an incinerator would produce '62,000 tonnes of ash, 5% of which at least would be hazardous, which was equal to 9,000 tonnes per annum'. This fly ash is 'heavily contaminated with toxins such as dioxins, furons and various heavy metals such as lead, cadmium and mercury'. The remaining ash, known as bottom ash was also a risky by-product, according to GSE. 50,000 tonnes of this ash would have to be landfilled in East Galway, thus negating one of the supposed benefits of incineration, in which it is claimed that incineration reduces landfill tonnage. At the meeting, GSE also claimed that a recent EU study entitled 'EU Dioxin and Health Data' linked exposure to dioxins with 'a range of health risks including liver damage, suppression of immune system, formation of cancers and abnormalities in foetal developments'. The EU report also detailed an up to 144 times higher risk of dioxin exposure to new born infants with incineration according to GSE.

5.2.2. Political Debates

GSE's campaign had an impact on local politics. Local media reports from the first two months of 2000 indicate the high level of public debate about the incineration issue at that time. In a three hours meeting of Galway City Council on 31 January, councillors expressed fears about the possible risks of emissions from incinerators. Consultant for M. C. O'Sullivan, P. J. Rudden, was present to answer the queries of the councillors on the issue. Concerns about potential health risks were expressed by some councillors, including Alderman Catherine Connolly (Labour) and Councillor Padraig McCormack (Fine Gael).

Rudden drew comparisons with the proposed incinerator and existing incineration plants in Europe, which were operating 'without problems'. When the scale of the plant was raised, Rudden referred to the fact that there were only two public objections, based on traffic concerns. However, the number of objections increased significantly as GSE's campaign gained momentum, and the deadline for public submissions was moved from March to May 2000 by the City Manager, who was responding to local politicians who were being informed by GSE's campaign. Some of the concerns voiced by councillors included a complaint from Deputy Padraig McCormack (FG) who criticised the lack of public debate on the issue, particularly on the fact that incineration was now part of a national waste policy. Alderman Catherine Connolly (Labour) was 'astounded' by the lack of scientific information presented to the councillors. In response, City Manager Joe Gavin said it was important to let the public know that the draft was available for inspection and submission. However, the original deadline left only a few weeks for this process to take place. Councillor Michael O hUiginn (Fianna Fail) echoed the sentiments of the government by saying Ireland was '40 years behind Europe' on the issue of waste treatment (*Galway Advertiser*, 3 February 2000).

On the Saturday previous to the city council meeting on 31 January, GSE had participated in the launch of the 'Galway Environmental Alliance', a network of anti-incinerator and anti-super-dump protestors. A public meeting in Eyre Square was followed by workshops in the Great Southern Hotel. These talks were attended by Michael D. Higgins, TD, as well as Councillors Connolly and Leahy. At this meeting, Aine Suttle of GSE argued that Greenpeace studies of Swedish and Austrian incinerators showed dioxin levels in ash by-products were much higher than official estimates. GSE were, according to Suttle 'concerned about where the incinerated ash will be disposed of, and in what manner'. Suttle also claimed that 'burning up to 120,000 tonnes of rubbish per annum is likely to produce up to 40,000 tonnes of toxic ash' (*Galway Advertiser*, 3 February 2000).

The focus of GSE was declared as being concerned with sharing information, monitoring topical issues and lending support to environmentally related projects. In response to concerns raised at this meeting, P. J. Rudden was described as 'lashing out' (*City Tribune*, 11 February 2000) at Aine Suttle's claims on toxic ash residues. Rudden accused Suttle of 'misleading the public by highlighting fears' over the amounts of toxic waste ash that will be produced through thermal treatment:

> Rudden described Suttle's statement as 'gross misinformation'. Rudden claimed that '30% of waste each year will residue as ash. Only 5% of that 30% ash is hazardous … which means that only 4,000 to 8,000 tonnes to toxic ash would be produced each year'. And that the toxic ash produced 'would more than likely be exported to a proposed special landfill site in either Dublin or Cork'. (*Galway City Tribune*, 11 February 2000)

At this point, some political response to the debate on incineration, and to GSE's campaigns could be detected. Speaking to the *Galway City Tribune*, 11 February 2000, Senator Margaret Cox (Fianna Fail), who also held a seat on Galway City Council, called for any decision on the incinerator to be put back nationally, to allow the reduction and recycling option methods of waste management to be tested. 'The public is roundly condemning the introduction of thermal treatment, (incineration) but the entire implications of it as a method of waste management were not being researched' (*Galway City Tribune*, 11 February 2000).

GSE's framing strategies had a major impact at every level of the debate on incineration. The local print media had covered the debate in an even manner, providing equal coverage for GSE's views on both the consultants' statements and subsequent political responses. This was due in part to GSE's ability to frame the debate in a manner that provided the local media with the type of source information on the health risks posed by incineration, as well as MC O'Sullivan's neglect in regard to providing accounts of these potential health risks. This framing of health risks was later picked up by local politicians, and in turn by the Galway newspapers. An example of this is provided in the 11 February 2000 edition of the *Connacht Tribune*, which included the following extract from a large article on the Waste Management debate

> Fears have been expressed over the toxic ash because it would contain amounts of dioxins; which are known to lead to a range of health risks including cancer, liver damage and damaging foetal development. In fact animal testing has shown that just 1 gram of dioxin is enough to kill 10,000 people. Greenpeace in Sweden recently reported that the amount of dioxins from incinerator ash ranged between 18 and 216 grammes per year. The Swedish Environmental Agency put the figure at between 3.6 and 5.4 grammes. (*Galway City Tribune*, 11 February 2000)

The statistics introduced in this article are congruent with GSE's interest led data, indicating that the local press were affording this 'interest driven science' some respect. With the information battle between both the pro and anti-lobby now heightening, both GSE and the City Council announced that they would hold separate public meetings to address the incineration issue further.

The first of these meetings was held by GSE on 7 February 2000. At this meeting, GSE attempted to highlight the health risks posed by incineration, and challenged the 'unstable alignments' that existed between the consultants and the local authorities. The waste plan consultants, M. C. O'Sullivan, were targeted by GSE for failing to provide information on the potential risks to health. GSE also accused Galway Corporation of 'not making it easy to get the information involving the incinerator plant' (*Galway Advertiser*, 10 February 2000). GSE had requested a website to provide information on the incinerator, but this had not happened. The consultants were also accused of not providing the public with enough information to the side effects of incineration. 'In 120 pages of information on the consultant's research, not once is the word dioxin used' (*Galway Advertiser*, 10 February 2000).

GSE used the February meeting to highlight the health risks to the farming sector that incineration could pose due to the processes by which dioxins lodged in animals, in a way that couldn't be treated. GSE had hoped to open up their movement to the agricultural sector, attempting to mobilise that section of the public that would also have concerns about health and economic risks. To do so, GSE also detailed the plight of French and Italian farmers who lost supply contracts with the Danone dairy company, who refuse to take milk from farms located near incinerators.

GSE highlighted a further area of environmental concern that could arise from the heavy truck movements required to dispose of the fly ash residue. According to GSE:

> If the plant is to burn 200,000 tonnes of waste, that leaves approximately 66,000 tonnes of ash. This would indicate that 10,000 truckloads of toxic ash would have to be transported from the plant each year. (*Galway Advertiser*, 10 February, 2000)

GSE's framing process utilised interest-driven material from Ireland and abroad. GSE referred to Dutch legislation prohibiting the use of fly ash without monitoring. It can never be placed near water, and if used for road foundations, it must be lifted and stored during repairs. As the ash will remain toxic for over 200 years, health implications were such that incineration had been condemned by the Irish Doctor's Environmental

Association, according to GSE. An alternative form of waste management, which avoided health risks, was available through recycling. To support this point, a video of a successful recycling scheme, with 60–70% recovery in four years in Ontario, Canada was presented.

The link with the campaign in Wexford provided GSE with their guest speaker, Joe Bridges who had been involved in the campaign against incineration in the South East, and was 'instrumental in having the proposed plant there rejected by the corporation' (*Galway Advertiser*, 10 February 2000). He also cited the difficulties facing the incineration industry in France, where three plants had been forced to close due to unsafe levels of dioxins on farms in their locality. Bridges also claimed that food companies in the South East have been asked by their European customers whether their plants or their suppliers are located within 40 miles of an incinerator plant.

One interesting aspect of the campaign's political opportunity structure that emerged from this meeting was the increasing involvement of politicians and concerned citizens from County Galway. GSE seized on the opportunity to maximise the link between the rural based anti-landfill groups and argued that the incinerator posed a risk to county residents due to the '50,000 tonnes' (of fly ash), 'as it is planned to landfill it in East Galway'(*Galway City Tribune*, 18 February 2000). This point led to Oranmore P. D. Councillor Noel Grealish inviting GSE to address the County Council on the issue at their next meeting, while the Galway County Manager, Donal O'Donaghue 'gave an undertaking to contact the City Manager over the setting up of a special forum on the incinerator issue' (*ibid.*).

GSE were able to create 'increased access' from their association with these 'influential allies'. Having gained support from political and public figures, GSE were now keen to hold further public meetings. The next public meeting held by GSE was set for Monday, 28 February. Under a front page headline declaring, 'Move to delay deadline on incineration decision', *The Connacht Sentinel* highlighted the possible health risks:

> Growing concern over the health implications posed by a proposed waste thermal treatment plant for Galway has prompted a number of local politicians to press for an extension of the proposed closing date for submissions on the plan. (*Connacht Sentinel*, 29 February, 2000)

GSE had made a point of contacting and inviting representatives of the city's various residence associations. Once the health risks of incineration were made apparent to these groups, local politicians were contacted, and the groundswell of public opinion began to turn not only against incineration itself, but also against the local authority, the state and their technocratic consultants, who had not been forthright on the health issue in the CWP. The meeting held at the Corrib Great Southern Hotel, also allowed GSE to press local councillors to extend the deadline for public submissions on the CWP beyond the 31 March 2000.

The siting of the incinerator plant also provided an opportunity for campaign mobilisation. The proposed sites, including Castlegar, Carrowbrowne and Galway Harbour, were near or in the city. Many speakers at the meeting 'expressed concern that an incinerator was proposed for one of four sites around Galway City at a time when there were genuine concerns about the threat to health posed by dioxins from the plant' (*Connacht Sentinel*, 29 February 2000). The meeting, which lasted four hours and which was attended by over 200 people, heard various City Councillors, from both government and opposition parties, support GSE's proposal to extend the submissions deadline. Deputy Padraig McCormack spoke of his intention to seek clarification on the waste plan. 'Everybody is speaking tonight though this was a "fait accompli", but you can take it that public representatives will educate themselves fully about this plan' (*Connacht Sentinel*, 27 February 2000).

5.3. CONCLUSION

As GSE established their health risks frame through their presentation of interest-led data, they were simultaneously extending their NIMBY frame in a manner followed by ecopopulists who utilise expertise and networks to establish wider campaigns. This progression was achieved through the mobilisation of key committee members in various roles most suited to their own expertise. This range of expertise was used to establish the health risks frame, as their issue-led data were highlighted through their media and public protest campaigns, a pamphlet and their Internet site.

GSE also extended their NIMBY frame by linking with the anti-superdump groups, most notably through their participation in the Galway Safe Waste Alliance. GSE put forward an alternative to both incineration and landfill in recycling, highlighting the economic benefits of recycling over the other two processes. In so doing, GSE were able to promote its own ecomodern alternative of recycling in contrast to another ecomodern approach, incineration, and the least favoured option of landfill. By framing the health risks and NIMBY issues in this way, GSE established not just localised opposition to the single issue of incineration, they also provided an alternative process in recycling, as well as extending their campaign to the anti-landfill groups. GSE's campaign had progressed from its NIMBY inception into an ecopopulist movement by exploiting the political opportunity of the CWP's omission of health risks and the local councillor's populist concerns. In the next chapter, GSE's emergence from the first two frames into a third frame of democratic deficit will be explored, in the context of the state's response to GSE's campaign.

This opportunity to mobilise the grievances of a concerned public, which in turn brought pressure to bear on local political structures, proved to be successful tactics for GSE, in the early phase of their campaign. By framing the debate around the potential health risks posed by incineration, GSE ensured a widespread reaction following their public meetings and media campaign. This strategy provided the group with two favourable outcomes. In the first instance, their attempts to mobilise widespread public support had been achieved. Secondly, by mobilising expertise on health risks and pointing to the neglect of these risks by the proponents of incineration, GSE were able to highlight the very health risks that the consultants had wished to obscure, which highlighted the superiority and independence of their expertise.

This chapter has examined the internal mobilisation phase of GSE's campaign. In particular, the chapter has set out GSE's first two frames of health risks and NIMBYism. These two frames were established in order to facilitate an interest led science challenge to the technocratic 'official' science of the state. GSE were able to achieve this form of access to the political structures of the state's waste policy by exploiting the state's own omission of health risks in the CWP. GSE were able to construct their frames from the mistrust of technocratic science established in the first phase of environmental campaigning in the 1970s and 1980s.

As such, the dispute about incineration became a contestation of two eco-modern approaches to waste management. On the one hand, the state supported the technocratic option for incinerators, while GSE's preference was for increased recycling in place of either incineration or landfill. By extending their health risks and NIMBY frames, GSE opened up a third frame, democratic deficit, which will be examined in the next section.

CHAPTER 6

THE DEMOCRATIC DEFICIT FRAME

6.1. INTRODUCTION

By extending their campaign beyond its NIMBY and health risk frames, GSE were able to open up a third frame, which was established from the political opportunity of democratic deficit surrounding the state's response to their campaign. This democratic deficit frame had its inception in the state's initial omission of the health risks of incineration in the CWP. By exploiting this opportunity, GSE were able to establish their own credentials through their provision of interest-led science about the health risks. Another area of democratic deficit exploited by GSE was seen in the state's attempt to rush the CWP through without undergoing the proper public consultation process that was written in to regional waste plans. Furthermore, the state's removal of the councillor's powers to decide the waste issue contributed significantly to a local sense of democratic deficit.

As the state and local authorities responded to GSE's campaign by closing the Poolboy Ballinasloe landfill before it had reached its capacity, GSE were able to argue that the state was in some way fabricating an immediate waste crisis to promote their regional waste policy. GSE also had to contend with the state's denunciation of the GSE campaign as being merely self interested NIMBYism based on 'scaremongering' tactics. However, the state's removal of the councillors' powers on waste management provided a further political opportunity for the framing of the democratic deficit issue.

6.2. LOCAL CONCERNS ABOUT DEMOCRATIC DEFICIT

The Menlo Park Hotel meeting also provided an opportunity to influence local political representatives. Catherine Connolly of the Labour party enquired about a proposed series of public meetings that the City Council had promised on the issue, but that had never taken place. Connolly also raised the omission of health risks from the consultant's report:

> It is a one-sided type of information we are getting about this plan and I think it is unequal this plan and I think it is unequal that a voluntary group such as GSE has to educate us about the other side of the plan. I want to thank GSE for all the information they have given me and the other councillors. (C. Connolly, *Connacht Sentinel*, 2 February 2000)

By distancing herself from City Hall on this point, Connolly became an 'influential ally' (Tarrow, 1994, p. 88) on the issue: Connolly had quickly gained a reputation for highlighting City Council's attempts to mislead the public on many issues, particularity in relation to Planning Law violation. By allying herself with GSE, Connolly was providing the resource of authority to the protest group. The Labour Party in Galway West had been supportive of GSE, with TD Michael D. Higgins meeting with the group, and offering information on tactics and dealing with the media (M. D. Higgins, interview, 20 November 2000).

Another strategic success for GSE was the announcement of an extension of the dealing for public submissions to the Waste Plan, from March to May. The decision by the Corporation to extend the submissions deadline was announced by the City Manager, Joe Gavin on 6 March 2000. Mr. Gavin stated that

> The reason for the extension of the deadline is to give everybody the chance to study as comprehensively as possible all aspects of the issue. We are inviting GSE to submit all available data they have for consideration and we will make available to them all the data that we have on the issue. (*Connacht City Tribune*, 10 March 2000)

It is clear from the City Manager's statement that GSE were winning the debate on the Waste Plan and were utilising the scientific information available to them in an effective way. GSE's ability to highlight the omissions of the health risks involved also provided an opportunity to control the direction of the debate overall. This can be contrasted with the narrow focus of the Local Authority and the City Manager, who appeared to want to rush the decision on the Waste Plan through, without either proper public consultation on the presentation of all the facts on the issue. Essentially, GSE were now in a position where many local politicians, the general public, and crucially, the local media, not only trusted the information it was providing but were relying on it for an overall understanding of the pros and cons of incineration. By exploiting this mistrust, GSE were able to construct their democratic deficit frame, developing their campaign from its single issue and NIMBYist inception.

The response of the authorities to the issue was portrayed as undemocratic by GSE, who seized upon the opportunity to frame the issue in a political manner. This allowed a further opportunity for GSE to portray the authorities as being unreasonable and misleading on the issue. Further pressure was applied when the city manager revealed that a decision would be made on the landfill issue by the end of the year. Ballinasloe was the only landfill site left in the county so there was an onus on the local authorities to make a decision on the issue (Joe Gavin, *Connacht City Tribune*, 10 March 2000). Clearly, the need to deal with the landfill issue was apparent to all participants in the waste management debate. However, by introducing a time frame into the debate, national and regional government were constructing a case for local authorities to rush through their decision in relation to the regional waste plans. Campaigners against incineration were portrayed by the authorities as having a NIMBY perspective, with self-interested individuals or groups preventing reasonable outcomes on the waste issue. However, the accusation of NIMBYism directed at GSE, can be disputed. According to Rabe (1994) 'a common NIMBY outcome' usually revolves around individuals with a grievance seeking compensation. GSE resisted seeking financial compensation, or legal challenges staying instead within the established structures of the campaign, where the political opportunities presented could be utilised for the benefit of a media and public mobilisation strategy campaign. By adhering to their established tactical approach, GSE hoped to maximise access to the POS, by maintaining their challenge to the CWP. In so doing, GSE's campaign had been transformed from a NIMBY dispute into an ecopopulist political campaign.

6.3. POLITICAL OPPORTUNITIES AND DEMOCRATIC DEFICIT

The campaign undertaken by GSE had challenged the state's waste policy, contesting the technocratic science of the state's incineration option, while also portraying the Fianna Fail/PD government as being undemocratic in their response to their campaign; to gain the ascendancy in the dispute after the CWP had been rejected and with Fianna Fail hoping to take three seats in the Galway West Constituency in the next general election, the state responded to GSE's campaign with a strong defence of the CWP.

At this point in time, there were no structures in place at the local government level to facilitate a public forum or committee that GSE could take part in or be appointed to. The state had established its waste policy on the basis of its own interpretation of the EU's waste hierarchy and had sent mixed signals on the introduction of municipal recycling. This ambivalence in their policy created the opportunity for GSE's challenge. In this way, GSE could achieve access to the political opportunities that emerged from the shifts in the state's waste management policy over the duration of their campaign, particularly where local authority councillors were taking votes on the CWP. The state's confrontative response to GSE created a further opportunity to create GSE's democratic deficit frame.

Whereas GSE's campaign didn't become 'radical' per se the state's response did allow GSE to shift the focus of their campaign, from a single environmental health, to a more political focus on the 'democratic deficit', allowing the state to overrule the concerns of communities in favour of their existing policy.

The state's defence of the CWP took place throughout the winter months of 2000 and 2001. While local councillors from the government parties had not defended the CWP in such a way that would have prevented GSE from wider mobilisation, ministers such as Frank Fahy of Fianna Fail and Bobby Molloy of the PD's brought out statements that were critical of GSE and their data on the health risks of incineration. These public condemnations were issued despite the support GSE had gotten from the public with their petition. In time, a waste crisis emerged on the streets of Galway, as the Ballinasloe landfill closed and uncollected rubbish piled up on the streets of Galway. This waste crisis, which may have been partially created by Minister Dempsey's initial refusal to renew the landfill's license, was seized upon by the state as evidence that the CWP needed to be introduced without further debate. However, the state's response to GSE provided further opportunities for GSE to capitalise on the democratic deficit angle, as they portrayed this removal of local authority power as undemocratic. The following sections examine the key events surrounding the state's response to GSE.

6.3.1. The State's Response to the Rejection of the CWP

In the Spring of 2000, Galway city council voted to reject the CWP. While GSE's campaign succeeded in influencing the local authorities in their rejection of the CWP, a response from government to this rejection soon followed. Galway City council's rejection of the CWP had put the waste management plans for all of the regions in jeopardy. The response to the council's rejection of the CWP was decided by the Minister for Environment and led to government spokespersons attributing blame for the waste crisis to individual GSE committee members. This strategy of blame had been used with dramatic effect during the Galway Bay sewage treatment plant dispute, when the spokesperson of the 'Save Galway Bay' campaign, Ray McBride was left with costs of over £160,000 after a High Court hearing ruled against him. At the first meeting of Galway County Council in September 2000, County Manager Donal O'Donohue stated that he had 'an executive duty' to put a waste plan in place, despite the county councillors' rejection of the plan. This statement was indicative of a tactic adopted later in the dispute by government when it removed the councillors' power of decision on waste management issues. Councillors from Fianna Fail and the PDs also began to castigate their city council colleagues, for 'putting a spanner in the works' of the regional waste plan. (*City Tribune*, 8 September 2000). At the meeting, it was also decided that the county council couldn't make plans for waste management in light of their city colleagues' rejection of the CWP. Councillors were limited in their options, as the EU's waste hierarchy had placed landfill as the least favoured option. GSE highlighted the fact that it was waste reduction and recycling that had been put at the top of the EU's hierarchy.

As the EU promoted eco-modern approaches to waste management, the populism of the Irish political system came to be superseded by a technocratic fixation with the incineration option. GSE's ecopopulism encouraged recycling as their alternative to both incineration and landfill, but while the state was theoretically in favour of some recycling, it never pursued that option to the extent it had promoted incineration. The option of landfilling was unpopular from the top down perspective of the EU and the bottom up perspective of the NIMBY groups. This dualistic pressure reduced the room for manoeuvre in political circle, particularly for local councillors who were faced with populist and electoral pressure on the issue.

As a result of these external pressures, councillors had indicated their reluctance to agree a new site for landfill in the county if it had to deal with all of the region's waste. Some councillors argued that there should be separate plans for city and county. Reports on the meeting included a warning from the county manager about the impending waste crisis and his executive duty to take a decision on the plan. The consultants for the CWP, M.C. O'Sullivan indicated that costs were rising. 'Landfill cost alone have risen from £5 (€6.35) a tonne to between £40 (€50.79) and £50 (€63.49) a tonne, and householders could be facing a hike in refuse charges in years to come' (P. J. Rudden, Galway County Council Meeting, 5 September 2000).

The emphasis on the rising costs of landfill was in line with the EU's waste hierarchy, which had landfill as the 'least favoured' option. However, the state had yet to introduce any of its proposed recycling schemes.

In mid-September, Roscommon County Council also rejected the CWP due to health fears about incineration. Local Fianna Fail councillor Michael Finneran was quoted as saying the CWP was 'in tatters' (*Galway Advertiser*, 28 September 2000). However, when Mayo County Council joined Sligo and Leitrim in accepting the CWP, it was reported that:

> Some Mayo councillors believed voting against the plan was futile as its contents can be imposed by the Minister for Environment or any authority that rejects it. (*Galway Advertiser*, 19 October 2000).

This indicates the pressure that was being put on local authorities by the state, who were threatening to override local waste decisions, and revoke local authority powers to make decisions on regional plans. A legal review of the city council's decision to reject the CWP was announced on 5 October 2000. The councillors only discovered about this review from a report in the *Irish Times*. The city manager referred the decision to see if it complied with legal requirements in light of the councillors' rejection of the plan. Further pressure was being applied by government sources by announcing that householders would face a large increase in refuse charges due to the city council's rejection of the CWP. The costs of after-care treatment at the Poolboy landfill in Ballinasloe were to be met by Ballinasloe residents according to the reports in the press. This announcement also served to increase the tensions between urban and rural councillors. Ballinasloe councillors were particularly annoyed that people living 40 miles away in Galway city were paying far less for what they see as their 'highly-subsidised refuse service' than residents of the East Galway 'capital', despite the fact that 'Ballinasloe people have to live within a mile of the landfill' (*City Tribune*, 20 October 2000).

By November 2000, both councils were beginning to put together a joint plan on recycling and landfill, in an attempt to resolve the impasse on waste management in the region. Both councils set up sub-committees to establish a joint plan for the county. At this time the legal report into Galway City Council's rejection of the CWP was presented to the councillors. It was criticised by Fine Gael's Padraig McCormack for merely rehashing the need for a waste plan, something councillors were already aware of:

> The legal report constituted a waste of time on the corporation's past when they should have been looking at the alternatives to incineration according to Deputy McCormack. (*City Tribune*, 7 November 2000).

A pilot project for recycling of domestic waste also began in November 2000. Some 664 houses in the Renmore area were chosen by the city council to gauge the potential success rate for any city-wide recycling scheme. The Renmore area was selected due to its good record on environmental community initiatives. The local residents' association was quoted as saying it was 'determined to use recycling to its utmost potential' (*Galway Advertiser*, 12 October 2000). A depot for recyclable articles was opened at the Sandy Road Corporation depot to facilitate the project, and a hazardous waste service was also provided, for items such as batteries. The segregation of household waste was divided into biodegradable and non-biodegradable, with separate 'wheelie-bins' being provided for each household by the council.

The city manager, Joe Gavin felt that rather than protesting a decision of the city council, people should lobby their local councillor and allow them to make any decisions on such issues as the CWP. While he acknowledged GSE's research informed the waste plan debate, he called for the group to release all of its interest-driven information to the council, rather than continue with their campaign against incineration. The technocrats hired by the local authority provided adequate information for the plan, according to Gavin. He claimed that a 'zero-waste' policy was 'too idealistic' and that by promoting this option, GSE were 'not coming forward with a practical solution' (*ibid.*). The need for a decision to be made 'quickly,' was also reiterated. He felt that the recycling targets set out in the Plan were 'very ambitious' (former City Manager, Joe Gavin, interview, 6 July 2000). According to Gavin, 'some lobby groups have the effect of delaying very important environmental projects, and can be seen as having a negative impact' (*ibid.*). Here, Gavin was highlighting the rising costs that would result from any delay in implementing the CWP due to GSE's campaign. Earlier this argument was used to turn public opinion against the protesters in the Galway Bay sewage treatment plant dispute. This 'negative impact' was a direct reference to the costs incurred by the delays surrounding the sewage treatment plant, which increased as that dispute dragged on. The city manager's comments were indicative of the authorities' fixation with incineration as an eco-modern technological solution, rather than recycling.

6.3.2. Manoeuvring Around the Landfill Issue

The state began to highlight their own arguments to support their waste policy, while criticising GSE's campaign. The issue of rising waste costs as a result of GSE's campaign was also used by the authorities in relation to the issue of the closure of the Ballinasloe landfill at Poolboy. The waste crisis deepened as Ballinasloe Urban District council imposed a 50% increase in charges for the use of the Poolboy landfill in Ballinasloe, County Galway. This led to reports of a new crisis for Galway City householders, who would have to pay a large increase in refuse charges. The problem was heightened by a complete ban on commercial waste at the site. Private firms, such as Barna Waste, were already collecting commercial waste in the city at that stage, had only three months to find alternative sites following the decision. Reports suggested that landfills as far away as Dublin, or even Northern Ireland, would have to be used as alternative sites for Galway's commercial waste. The city's business community were now angered at being faced with a significant increase in waste charges for their recently privatised collection service.

GSE's interpretation of the state's response to their campaign was that they were creating a sense of crisis on the waste issue, turning the councillor's decision to reject the CWP to set urban and rural communities at odds over what alternatives could be pursued. The waste 'crisis' could have been resolved by the state extending the lifetime of the landfill at Ballinasloe, which wasn't initially due for closure until 2005, at existing cost, while alternative recycling and improved landfilling projects were implemented.

However, it could be said that by allowing both a crisis to develop in addition to a degree of opposition between city and county councils over the issue, the state was attempting to force the city councillors to review their decision, while ultimately placing blame for the crisis on GSE and their supporters. By attempting to demonstrate the hazards of not implementing existing waste policy, the state was attempting a strategic manoeuvre to counter GSE's success in the first phase of the campaign that set urban and rural areas at odds over what alternatives could be pursued whereas the 'crisis' could have been quickly resolved if the state extended the licence of the Poolboy landfill.

The state's response to the GSE challenge could be seen as a manipulation of the waste issue, to create a sense of immediate crisis among the commercial sector and to discredit both local politicians and community groups who had opposed the regional waste plans. The pressure on city and county councils continued when the Department of the Environment sent formal letters to Galway Corporation and Galway County Council warning them of the need to move urgently on implementing waste policy, and stating that Galway couldn't opt out of national waste policy plans. The letter, which was described as 'a snorter' by the local press, was said to be 'uncompromising' in its tone:

> I, the Secretary General of the Department of Environment, write to express our serious concerns regarding the present position in Galway in relation to the proposed waste management plan for the Connacht region ... The overall waste management planning process has been underway for several years and its completion is now seriously overdue. As you know, the European Commission has initiated legal proceedings against Ireland in the European Court of Justice on grounds of non-compliance with EU requirements concerning waste plans. Accordingly, from a national perspective, it is imperative that the process is satisfactorily concluded within the shortest possible timescale.
>
> It is most undesirable that one authority would, in effect, seek to withdraw from a well established planning process at a very late stage, thereby negating the efforts and legitimate expectations of other authorities concerned as well as threatening the overall integrity of the regional plan. It is difficult to see how Galway acting on its own could hope realistically to attain landfill diversion and waste recycling objectives equivalent to those in the proposed regional plan. Accordingly, in my view it is very important that Galway Corporation take the opportunity to reconsider, as soon as possible, its position regarding the proposed regional plan, having regard, to the consideration outlined above. (*City Tribune*, 24 November 2000)

What were evident from the letter were the attempts by the state to pressure members of Galway's councils to reverse their decisions and accept the CWP. This somewhat undemocratic approach to the elected councils' decision began to cause as much concern to GSE members as the health risks posed by incineration itself. The question being asked by GSE activists was 'why do the state seem to insist on introducing incineration, when there are cheaper and safer alternatives?' The issue of democratic deficit was now being incorporated into GSE's framing of this dispute.

The councillors discussed the state's response to their rejection at a meeting in November 2000. The meeting heard claims by Catherine Connolly that the Fianna Fail/PD government was insisting on going ahead with incineration, while government party councillors launched attacks on their opposition colleagues. PD councillor, Declan McDonnell warned that the situation could emerge in the not too distant future when they 'might be forced to stop taking the domestic waste from the city'. Fianna Fail councillor, Val Hanley, informed the meeting of the need for a market-led strategy, as household waste charges could increase to up to £300 per house, at a time when the price per tonne for waste cans and paper was falling. The remarks of the government party councillors indicated the pressure they were being put under by their party organisation to create a sense of crisis over the councillor's decision to reject the CWP. The issue would be revisited after the turn of the year, when the county council would have decided on an alternative landfill site.

The sense of crisis surrounding the waste management issue was heightened by the announcement of increases of up to £50 (€63.49) per year in domestic refuse charges. Business in the city had to pay an increase of 8% in their waste charges. The increase was attributed to increasing costs at Ballinasloe dump, which was set to increase by 50% from £40 €50.79) per tonne in 2000 to £60 (€76.18) per tonne in 2001. Other costs, associated with labour and transport, were also factors in the increase. The total estimated charge would amount to £2.7million (€3.4million) during 2001.

The increase in waste charges also served to increase the pressure on Galway's citizens and their representatives to think again on the CWP. As the city's rejection of the plan had halted the government's regional waste plans, and the state was unwilling to defer from its 'dominant strategy' on waste management, the councillor's rejection of the CWP was unacceptable. This response to collective action has been defined as being, at times, more effective than repression (Tarrow, 1994, p. 95). As the state 'depresses' collective action through legal or political means, movements or individual activists can become isolated, with the result that collective action can dissipate.

While the state increased the pressure for a reversal of the rejection of the CWP, GSE moved to build on the success of the Renmore recycling scheme, calling for the introduction of recycling on a citywide basis. GSE criticised Galway Corporation for failing to fund any kerbside recycling schemes in the city. GSE argued that a recycling scheme had been offered to the entire city during the CWP debate. GSE claimed that 20,000 householders in Galway were being 'denied the chance to recycle' because of this lack of funding, despite the public's desire to participate in such a scheme. GSE also argued that householders who achieved high recycling rates should have a reduction in their waste charges, rather than the increase put forward by the corporation:

> It is unfair that householders who have invested in recycling bags and compost bins, and who in some cases achieved diversion rates of over 70%, are going to be asked to pay an increased waste charge of £170, the same as those who are not making any effort to recycle. (C. Ó Brádaigh, *Galway Advertiser*, 30 November 2000)

GSE also claimed that a recycling scheme would help the corporation to reduce its Book of Estimates costing of waste management. At the very least, the group felt the local authorities should encourage those who did try to recycle. GSE discussed this issue at a general meeting held on

28 November 2000. At this meeting, criticisms were made of the fact that the corporation had made no provision in the Book of Estimates for recycling, despite earlier promises to do so through the campaign. Reductions for those who recycled were also called for. The GSE speakers argued that the public were 'crying out' for recycling in the city. The meeting also discussed the need for grants to cover the costs of recycling bins and bags, compost bins and other materials, for householders who wanted to recycle and compost.

The state's attempt to raise the issue of costs was countered by GSE's focus on recycling as a viable alternative. GSE emphasised that their campaign was about the economic value of recycling. The increased costs of landfilling made recycling the most economically viable, according to the group. The groups argued that the 22,000 signatures of their petition against incineration wanted recycling as the alternative mainstay of any waste management strategy. GSE also announced at the meeting that they would launch a series of recycling awareness meetings across the city, throughout December, January and February. These meetings proved to be very successful, with over 100 people attending the first recycling meeting at the Castlegar Sports Complex on 11 December 2000. Local recycling businesses addressed these meetings alongside GSE committee members.

The state could no longer tolerate any more delays in the implementation of its required waste plans. GSE condemned the announcement by the Minister for the Environment, Noel Dempsey, that he would consider imposing waste plans on councils across the country that had rejected them. GSE saw this as 'a complete negation of democracy' (*City Tribune*, 1 December 2000). GSE pointed out that the large numbers that had signed their petition, together with the rejection of the CWP by the city's councillors, constituted a democratic decision on the issue, and one that shouldn't be overridden by the Minister. GSE also linked the lack of funding for recycling with the rejection of the CWP:

> If the Minister for Environment is withholding the funding from Galway Corporation that is needed to get city-wide recycling started, because they rejected the Connacht Waste Plan and incineration, then the people of Galway deserve to be told this. Financial support for recycling schemes from central government should not be conditioned on acceptance of incineration. (*City Tribune*, 1 December 2000)

The external pressure being brought to bear by EU requirements were also raised at this stage. The threat to overrule the decision of Galway City Council by the Minister was announced at the same time as a fresh attempt by CWP consultants M. C. O'Sullivan to use the threat of EU sanctions as a reason to force an overturning of the CWP vote. Their spokesperson, P. J. Rudden, claimed that the EU Packaging Directive could not be fulfilled without thermal treatment. 'Rudden warned that the waste crisis in Galway was mounting; he felt that the elected members were burying their heads in the sand on the issue' (*City Tribune*, 22 December 2000). Rudden further claimed that incineration was 'cheaper than landfill' (*ibid.*) despite his opponents focus on recycling as an alternative.

One piece of good news for GSE came with the announcement that the German incinerator 'held up as the model for Ireland' to copy was forced to shut down 'for spewing heavy metals over the German countryside' (*Galway Advertiser*, 16 November 2000). This thermo select gasification plant had been the subject of a visit by city councillors on a fact-finding mission the previous year. GSE welcomed the closure, and claimed it reinforced the need for an alternative waste plan for Galway. Gasification is a type of waste heating treatment that reduced waste to a gas vapour. It was considered as dangerous as incineration by GSE, as it also produced toxic emissions.

> Gasification was promoted as being the new emerging technology, and we can see it doesn't work. This is proof it has failed and was doing damage to the environment. We are calling for a new waste plan for Galway city and county and interested bodies such as the Chamber of Commerce, local environmental groups, and waste contractors should be involved as its better to do it right. (*Galway Advertiser*, 16 November 2000)

Economic factors continued to be highlighted by the state in response to GSE's campaign. The debate continued with a warning from the Industrial Development Authority that the continuing waste crisis could halt the economic growth of the city. The IDA predicted a 'doomsday' scenario where over 500 tonnes of waste would pile up outside the businesses, factories and offices of the city, due to Ballinasloe Urban District Council's decision to ban commercial waste from their landfill in October.

> The Industrial Development Authority this week warned that even Galway's major factories could face real problems. If these companies have no place to store their waste, then it will have to be stored on-site for a while, but that can only go so far. Ultimately, they may have to ship waste out of the country, which would come at a horrendous cost ... if these companies are unable to dispose of their waste, this will pose an unquestionable risk ... this could prove to have a hugely detrimental effect on these businesses, and on Galway as an attractive location for new industry. (IDA spokesperson, *City Tribune*, 22 December 2000)

The city's private waste contractor's association also commented on the looming crisis facing the city's commercial sector. The WCA was comprised of Walsh Waste, Barna Waste, The City Bin Company, Moran Waste, East

Galway Waste, Cappagh Waste, Kelly Waste and Corrib Waste, most of whom had major collection services around the city. The WCA said that all options for any proposed new landfill site on the island of Ireland had been investigated to no avail. They claimed that once collection vehicles became full, the companies would have to cease collections. It had also claimed that other local authorities would refuse to deal with Galway's waste. The WCA called on Ballinasloe UDC to reverse the decision on the Poolboy landfill. At a meeting with the employer's group IBEC, the private collectors were told by a senior Department of Environment official that 'the Minister may have to intervene in the matter and use his powers under the Waste Management Act 1996' (*City Tribune*, 12 January 2001).

6.3.3. The Galway City Waste Crisis

The first convening of Galway City Council in early 2001 occurred amid speculation that a second vote on the incineration proposal in the CWP would occur. The two sides of the city's waste 'crisis' are apparent in a press article discussing the meeting. The sense of impending crisis is presented in one paragraph that states:

> Now the waste issue is on the agenda for discussion at next week's Council meeting again and with the threat of intervention is the crisis by Minister Dempsey, speculation is rife that the councillors will be asked to vote again in favour of a common strategy between county and city .... (*City Tribune*, 22 December 2000)

This paragraph, which portrays the Minister as looming in the background and attempting to push through his policy, can be contrasted with the final section of the article, which puts the 'crisis' in perspective:

> It is understood that the Minister could use his powers to lift the cap on tonnage going into the Poolboy landfill site, which would remove the need for a ban on city waste. (*ibid.*)

One way of looking at the Poolboy closure would be that the government hoped to heighten the sense of crisis surrounding the Poolboy landfill closure in an attempt to pressure Galway's politicians into overturning their decision on the CWP. At the end of January, the city's newspapers carried headlines such as 'Rubbish May Pile Up on Streets as Dump Crisis Peaks' (*Connacht Sentinel*, 30 January 2001) and 'Galway to Become a Sea of Waste Unless Minister Acts' (*Galway Advertiser*, 1 February 2000).

The relevant articles presented a bleak picture. The *Connacht Sentinel* commented that an 'unprecedented waste crisis with hundreds of tonnes of rubbish was set to start piling up from Thursday morning next' (*Connacht Sentinel*, 30 January 2001), whilst the *Advertiser* noted how 'Galway city woke up this morning facing the possibility of being turned into a sea of waste over the next few days as the impasse over waste collection worsens' (*Galway Advertiser*, 1 February 2000). Undoubtedly, the city's politicians baulked at the sight of such headlines. The articles provided vivid warnings about rubbish piling up outside the shops and restaurants of Galway. The Galway Chamber of Commerce Public Relations Officer, Maeve Joyce, expressed the business sector's growing frustration; 'Is this a fair deal for the ratepayers of Galway?' (*Galway Advertiser*, 1 February 2000). The country's largest trades union, SIPTU, were quoted confirming that the jobs of the city's waste collector's were under threat, demonstrating that the unions were reflecting the stance of their neo-corporatist partners, the business lobby and the government.

The two councils held emergency meetings with the Waste Contractors Association, who expressed their concerns at being caught up in the dispute between local and national government. The waste collectors also warned that household waste would be left uncollected. They called on the Minister for the Environment to intervene. The minister's intentions were interpreted as follows:

> It now looks likely that Dempsey will use the current crisis to press Galway County Council and Corporation to adopt the Connacht Waste Plan, which was rejected by both authorities last year. (*Galway Advertiser*, 1 February 2000)

This statement was included with accounts of how the Minister could use his powers to extend the Ballinasloe landfill in a bid to prevent the crisis escalating.

GSE responded to these events by announcing a protest march and rally in February 2001. GSE committee member Peter Butler claimed that 'a secret deal was being hatched between Fianna Fail councillors in the city and county, and the Minister for the Environment, to overturn the unanimous votes of Galway Corporation and County Council against the Connacht Waste Plan' (*City Tribune*, 26 January 2001). GSE portrayed this 'deal' as the worst of both worlds that would include incineration and a superdump.

Galway Senator and Fianna Fail city councillor Margaret Cox accused GSE of 'scaremongering' and dismissed the claims by GSE of a secret deal as a 'total fabrication' (*ibid.*). Senator Cox revealed that she had met with Minister Dempsey to discuss the issue the previous week. She reiterated the Minister's call for the passing of a regional waste plan. Local media accounts of the suspension of waste collection continued to portray the sense of severe crisis now facing the city. Private waste contractors organised a picket on the Ballinasloe Poolboy landfill to highlight their plight. Press accounts spoke of fears that the ongoing crisis would affect the major multinationals in the city that produced large amounts of waste daily. The Irish Hotels Federation claimed the dispute could only harm Galway's tourist industry, as uncollected rubbish bags began to be stacked high outside of Galway's shop fronts. Images of rubbish bags accumulating on the streets were broadcast on the evening news, as part of a report on the crisis by RTÉ's Western Correspondent, Jim Fahey, in the last week of January 2001.

In the run up to GSE's public march on 3 February 2001, the group announced two new strategies in light of recent developments. One was the intention to seek a judicial review and the second was to make incineration an issue in the following year's general election, although this never took place. GSE also criticised the government for its handling of the waste crisis:

> The people of Galway are entitled to ask who these councillors actually represent on Galway Corporation ... we have said since last October that a waste crisis would be stage-managed to give political cover to Fianna Fail and the Progressive Democrats, to push this waste plan through for Galway. This is now what has happened. (*Connacht Sentinel*, 6 February 2001)

The march and rally called by GSE was scheduled to proceed from Fr. Griffin Road on the west of the city, on a route through the city's pedestrianised area, and on through to Eyre Square in the city centre. As the march was being held on Saturday, GSE hoped to maximise their potential exposure to, and support from, the Galway public. Scheduled speakers for the event included Labour TD Michael D. Higgins, Fine Gael's Padraig McCormack, and Fianna Fail county councillor Tim Rabbitt, who had supported GSE's anti-incineration stance from an early stage of the campaign.

The march was set for two days before the city council was to meet, on 4 February, to review its previous decision on the CWP. With an overall attendance of around 300 people, the march proceeded through the city. Among the groups supporting GSE's march were the anti-superdump groups, the Galway Safe Waste Alliance, and members of all strands of party politics, including Rabbitt of Fianna Fail. The march comprised of people of all ages, from students to pensioners. The participants were visibly supportive of GSE's anti-incinerator stance, waving their 'no incineration' placards and cheering GSE's spokesperson Conchúr Ó Brádaigh as he addressed the crowd. The rally was addressed by a number of local politicians, who reiterated their opposition to incineration. Repeated calls were made by the speakers to Galway city council not to reverse its decision to reject the CWP.

As the meeting unfolded, it became apparent that the government party councillors had bowed to pressure from Environment Minister Noel Dempsey, and would now vote in favour of the plan. Senator Margaret Cox, who broke FF party ranks to vote against the plan. Cox was later to controversially change her mind and vote for the CWP in Seanad Eireann. The councillors who opposed the plan argued that since a decision had already been taken against the plan six months previously, the new motion was similar to a Material Contravention, and would require 11 of the 15 councillors to vote for the waste plan. Media reaction to the council decision was intense. Headlines such as 'Corpo's U-turn on Incinerator' made banner headlines, accompanied by telling images of uncollected refuse sacks piling up on the city's streets. Reports on the meeting revealed its controversial nature:

Senator Cox defended her decision to break with the party line on the vote, saying:

> At no time were we told in Galway that 'waste to energy' ... was going to prove so controversial, as it had done in other countries. (Sen. Margaret Cox, *Irish Times*, 19 February 2001)

Senator Cox also believed it was unfair to expect councillors to decide on an issue 'without proper time or support it is often forgotten that councillors are part-time representatives, and waste is a hugely complex issue' (*ibid.*). GSE also expressed its 'extreme concern' about the 'U-turn' by the councillors. The initial advances of their campaign seemed to be lost, as the POS now began to close as a result of the councillors' U-turn. Conchúr Ó Brádaigh admitted that he thought the first vote would stand. An emergency general meeting of GSE's committee was called to discuss the new situation. And a public meeting was called for 27 February 2001. At this meeting a veterinary surgeon, Gerry Browne from Roscommon, spoke of the health risks posed by incineration for farm animals and discussed how dioxins enter the food chain. GSE committee member, Peter Butler, spoke about organic farming. The meeting was not as well attended as previous meetings, and there was a subdued atmosphere throughout. GSE's campaign was experiencing difficulties, as the morale of activists and committee members had been sapped by the reversal of the city council vote.

This sense of 'activist fatigue' corresponds with theories of cycles of protests (Della Porta and Diani 1995, Tarrow, 1994). As the strategies made by movements evolve, succeed or fail, activists may lose interest in the campaign. As campaigns tend to correspond with intensified activity combined with periods of revision and strategic regrouping, movements are vulnerable to periods of weakness and fatigue amongst participants. Such difficulties were being faced at this stage of the campaign by GSE.

6.3.4. Tactical Approaches to the Issue by the State

The tactical approach of the state built on the emergence of a waste crisis around the closure of the Ballinasloe Poolboy landfill. The problems with the landfill's capacity gave the authorities an opportunity to challenge GSE's successful campaign head on. By creating a sense of crisis around the Poolboy closure, the authorities were able to present what appeared to be a bleak scenario for Galway City, as uncollected waste piled up on the city's streets. As these images appeared on television and in the press, the spectre of further rate increases loomed over the city's commercial sector, already angry about the recent introduction of additional charges. Direct pressure was applied to the government parties' city councillors, including the unusual presence of Fianna Fail TD and Cabinet Minister Frank Fahey in the council chamber on the night of the vote to overturn the prior CWP decision.

The state stepped up its criticisms of GSE. Local Progressive Democrat Junior Minister, Bobby Molloy, criticised GSE's committee through the media, accusing them of 'scare mongering'. Under a banner headline announcing 'Molloy tears strips off 'scare mongering' Galway lobby groups', GSE's committee were 'lambasted' by the Department of the Environment Minister of State.

> We are seeing a continued campaign of scare mongering and innuendo. Selective and misleading information is being propagated by members of GSE in their opposition to the proposed Connacht regional waste plan. As a public representative, I must be scrupulous in my statement and in the presentation of facts, as I know them. GSE spokespersons, obviously do not feel the same obligation. (*City Tribune*, 9 March 2001)

At this point of the dispute, a hierarchy of individuals became pivotal to the state's aim of getting the CWP vote overturned. At the top of this hierarchy was Noel Dempsey, the Minister for the Environment. Dempsey's role was in highlighting Galway's rejection of the CWP as a problem for all of the country's regional waste plans, which he claimed had to be introduced together. Dempsey used Ireland's poor EU record and subsequent legal difficulties as a focus for putting councillors who opposed waste policy under pressure, while attempting to portray GSE as unreasonable opponents who were blocking good policy initiatives by the government. Junior Minister Bobby Molloy and Cabinet Minister Frank Fahey were both Galway based TD's, and they were both vocal in their targeting of GSE's committee during the campaign. Both were instrumental in applying pressure to their party colleagues at council level to reverse their rejection of the CWP.

At a local level, the role of City Manager Mr. Joe Gavin was also an important figure in the first phase of the campaign. Mr. Gavin repeatedly reminded the city councillors that they would be preventing the nationwide adoption of the regional waste plans if they rejected the CWP, as the regional plans were dependent on all councils accepting the initiative. The city manager was accused by Labour councillor Catherine Connolly of trying to force the CWP through, and of using his power to place other important items for debate, at the same time as the CWP was up for discussion. This left less time for debates, while the City Manager added further pressure when he warned of the need for a quick decision on the waste plan.

The ultimate tactic introduced by the state was the announcement by Minister Dempsey that planning decision on waste policy would be removed from the hands of the city's councillors. Dempsey wanted to introduce a Waste Amendment Bill that would place all power to make decisions on waste policy in the hands of the appointed city and county managers. The threat of removing the councillors' powers on the waste issue was repeated by the minister throughout the debate. GSE claimed the amendment would be 'a major assault' on local democracy.'

> Dempsey has been unable to convince the democratically-elected councillors. Incineration has never been put before the people and it's undemocratic to impose it. (*Galway Advertiser*, 8 March 2000).

The minister announced that his 'patience had run out' and that the Bill would be rushed through the Dáil and Seanad. 'The new Waste Management Bill should have passed through both houses of the Oireachtas by Easter' (*City Tribune*, 30 March 2001). While the dispute about incineration had been controversial, the Minister's attempts to withdraw councillors' powers on the issue drew criticism from mainstream political figures such as Connacht Ulster MEP Rosemary 'Dana' Scallon, as well as other Galway

TD's and politicians across Galway. Fine Gael Senator Fintan Coogan criticised the Bill in the local press:

> It's the heavy hand of centralism. This undermines the rights of local communities. (*Galway Advertiser*, 29 March 2001)

One result of the Minister's Amendment Bill was a second rejection of the CWP by Galway's county councillors, angered at the decision to remove their powers on the issue. The councillors drew up their own waste plan and planned a legal challenge to the Minister's Bill. The councillors' waste plan included a high percentage of recycling, a remote landfill, and no incineration. Fianna Fail councillor Matt Loughnane criticised the Minister's Bill, likening it to 'dictatorship', while Cllr. Paddy MacHugh said the Minister was 'obsessed by incineration' (*Galway Advertiser*, 12 April 2001). The perception of democratic deficit surrounding the Minister's Bill removing the councillor's authority on waste issues was now becoming as much of a concern as the health risks of incineration. Fine Gael TD Paul Connaughton said the legislation was

> the most draconian that had been drawn up for many years and the worst step for local democracy in a long time. (*City Tribune*, 10 April 2001)

At a time when GSE were facing a reversal of their initial campaign successes, reaction to the Waste Amendment Bill served to create a new wave of anger against Minister Dempsey and the regional waste plans. This frame created a new political opportunity for GSE. Strategically, the government may have overplayed its hand on the waste issue, as pressure on councillors gave way to the complete removal of their decision-making power. GSE seized on the county council's rejection of the CWP, claiming the first vote of the city council and the second vote of the county council as 'the way forward for Galway':

> In total, only 8 of the 45 elected councillors in Galway have supported his [Dempsey's] waste plan. To seek to implement the rejected Connacht Waste Management Plan in the face of such overwhelming rejection by the councillors is a recipe for public unrest. (*Galway Advertiser*, 12 April 2001)

The situation for local politicians hit a low after a High Court decision to fine Galway City Council £50,000 (€63,485) along with a threat to imprison the new city manager, John Tierney, for contempt of court over the reopening of the city's landfill at Carrowbrowne, which was already deemed full. Councillors reacted with dismay to the charges, as the issue seemed to become intractable. For the city and county councillors, the political opportunity structure was now completely restricted on this issue, as they had lost their decision-making powers. Owing to the clientelist nature of politics, councillors were always vulnerable to the populist pressure of groups such as GSE. However, they were even more vulnerable to the demands of central government in this regard. The issue was left unresolved through the summer adjournment of 2001, until the August reconvening of the council when only a month remained for councillors to make a decision on the CWP. If the councillors rejected the plan again, the city or county manager was now in a position to pass the waste plan over the councillor's heads.

However, on 12 September 2001, city manager John Tierney exercised his powers under Minister Dempsey's Waste Amendment Bill and adopted the CWP in full ignoring the amendment made by the councillors to omit incineration from the plan. The city manager's decision was met with criticism from the councillors. Alderman Catherine Connolly claimed 'this legislation undermines local democracy' and called for a condemnation of the Act, while Deputy McCormack complained that the Plan was being imposed from above:

> Giving the right to the manager to do that was a very contrite step. We have no say in what kind of Plan we have. This is not the Plan we wanted, but the Plan the Minister and the consultants wanted. I hope the city won't suffer in the long-term. (Deputy Padraig McCormack, *Connacht Sentinel*, 18 September 2001)

The councillor's protests came to a head after a visit by Minister Dempsey to Galway City Hall in August 2001, where the Minister announced the Waste Amendment Bill. The visit gave rise to the ironic image of the city's councillors greeting the Minister in full civic garb, as he arrived to remove their powers, while GSE staged a protest outside the gates. One exception to this was Alderman Catherine Connolly, who took her place among the protestors outside the Hall. At the 12 September 2001 meeting, Alderman Connolly proposed that the Minister be informed of the council's 'greatest disgust with what he has done to local democracy' (*ibid.*). Clearly, GSE's campaign against incineration had spilled over into mainstream politics, embroiling politicians from all parties, and had led to drastic measures being introduced by the government to curtail their successful opposition to incineration. As the 2002 General Election approached, the incinerator would become a key issue in the Galway West Constituency. The next chapter will explore the implications of this, and the impact the issue would have on the election outcome, as GSE came to expand their democratic deficit frame through electoral politics.

6.4. CONCLUSION

This chapter has explored the extension of GSE's campaign beyond their initial health risks and NIMBY frames into a third frame of democratic deficit. This third frame was established through the exploitation of the political opportunities of the state's omission of health risks in the CWP, and the portrayal of the waste crisis as the state's manoeuvring on the closure of the county landfill. The largest political opportunity for GSE to exploit was the state's removal of the Councillors' decision-making powers on waste management issues.

GSE were able to gain 'increased access' and gain leverage from these political opportunities due to the state's undemocratic response to GSE's campaign. While GSE had received 22,000 signatures to their petition, the state dismissed them as NIMBYists and 'scaremongers.' GSE were also put under pressure during the Galway City waste crisis, a period that saw the members of GSE come under sustained criticism from cabinet ministers while television images of uncollected rubbish piled high on the streets of Galway were broadcast.

GSE's campaign was coming under pressure from the authorities, but as the state removed the councillor's powers on waste issues, a further opportunity to present the democratic deficit frame was created. GSE were also able to rely on the support of 'influential allies' in the figure of councillor Catherine Connolly, who stood with them at their protest outside city hall as the minister arrived to remove the councillor's powers on waste matters, as well as local county councillors. By creating such political alliances, GSE were able to politicise their campaign, something that would allow them to extend the democratic deficit frame during the political opportunity of the 2002 general election.

CHAPTER 7

THE OPPORTUNITIES AND CONSTRAINTS OF THE DEMOCRATIC DEFICIT FRAME IN THE 2002 GENERAL ELECTION

7.1. INTRODUCTION: THE SHIFTING DYNAMIC OF POLITICAL OPPORTUNITY STRUCTURES

This chapter looks at the shifting political opportunity structure surrounding the 2002 General Election and the emergence of the democratic deficit frame. Over time, GSE's campaign had moved from being a single-issue stance against incineration into an ecopopulist campaign that had gained access to the political opportunity structure of the general election campaign. The election can be seen as a key external resource for GSE, in their attempts to 'increase access' to the formal political process. GSE were now in a position to bring their mobilised public support to bear on the political process, through the opportunity of the 2002 election campaign. To illustrate the constant shifts that occurred at this point in the campaign, a series of episodes will be examined. These illustrate the changing nature of the political opportunity structure that GSE brought as the incineration issue into the electoral campaign. The critical events of this phase of the case are as follows:

1. The Menlo Park Hotel meeting of GSE, on 13 January 2002, when a response to the overturning of the Council's rejection of the CWP was made.
2. The GSE Emergency General Meeting (EGM), in February, when GSE decided to make incineration an issue in the General Election. This led to tensions, over whether GSE should back only a Green Party candidate in

the general election, or whether they should also support politicians such as the PD's Noel Grealish who had campaigned with them.

3. The debate on incineration between election candidates organised by GSE in April 2002. All candidates appeared at the meeting, demonstrating the success of GSE in making incineration an election issue. Senator Cox came under pressure during this debate, particularly from PD candidate Noel Grealish.
4. The General Election campaign, throughout April and May 2002. By this time, all parties, except Fianna Fail, had adopted policies consistent with GSE's call for a rejection of mass-burn incineration in their election manifestos.
5. The 2002 General Election outcome, a reversal of fortunes in regard to the political opportunity structure. While GSE's anti-Fianna Fail campaign may have had some part in preventing Senator Cox from gaining a seat, their support for successful candidate Noel Grealish of the PDs, was later constrained by the returning coalition's government's programme, which included incinerators. The demands of coalition government would effectively prevent Deputy Grealish from following through on his promise to support GSE's stance in the Dáil, due to the agreement between the parties in their Programme for Government. The 2002 Programme for Government states that the FF/PD government have come out against 'mass-burn incineration of unsegregated waste, with no energy recovery, describing it as 'not an acceptable practice today'. It also states that the best available technology 'must be based on prior extraction from the waste stream of recyclables and problematic materials such as metals and batteries' to the maximum extent possible' (2002 Coalition Programme for Government). This position corresponds with the PD election manifesto, rather than with that of Fianna Fail, but this is as far as GSE or Grealish could get in their access to the political structures surrounding the coalition's waste policy.

For their part, GSE went into 2002 with a vision of how they saw the year unfolding. 2002 would see the holding of a general election, and the momentum of GSE's petition gathering and public meetings could be utilised in an attempt to politicise the incineration issue as part of the overall election campaign. The shifting dynamic of the political opportunity structure surrounding the incineration issue had seen GSE and their local political allies losing momentum, due to the removal of the councillors' powers on the waste issue. However, the response of a public now concerned at this perceived undermining of democracy allowed GSE to extend their democratic deficit frame.

This transition from the initial frames of health risks and NIMBYism to the democratic deficit frame led to some discord amongst GSE committee members unsure about the new political direction of the campaign. While many felt that someone from GSE's committee with political experience such as Conchúr Ó Brádaigh would make a good election candidate, some of the underlying tensions between committee members would surface throughout the election campaign. The year would start on a positive note for the group, as Dr. Paul Connett led the speakers at a GSE public meeting at the Menlo Park Hotel in January, Environment Minister Noel Dempsey, being the target of GSE's criticisms about democratic deficit. GSE members were concerned that a single-issue anti-incineration candidate might not be electable or that such a candidate might prevent a sympathetic candidate from being elected. By February, none of the leadership had made clear their intention to run in the campaign, so an 'influential ally' was sought from the local political scene.

GSE decided to call for all parties to reject incineration in their manifestos, and backed first the Green's Niall Ó Brolcháin, from the opposition, and later the PD's Noel Grealish, in an attempt to ensure they had, at least, one sympathetic voice at the cabinet table. This tactic was to backfire, as the political demands of the new coalition government, comprised of Fianna Fail and the PD's was to supersede any local concerns about incineration.

The populist electoral considerations of the parties created further political opportunities for GSE during the election campaign. In the Galway West Constituency, Fianna Fail had targeted three seats for the 2002 general election. The two sitting TDs in the constituency, Eamon Ó Cuív and Frank Fahey, hoped that they would be joined by Senator Margaret Cox. Cox proved to be an interesting figure during the campaign, speaking first against incineration, but then voting for it in the Seanad.

Political parties' electoral hopes are reliant on the populist Irish political system. This populist system has shaped a great deal of the political debate around incineration. While Environment Minister Noel Dempsey removed Local Authority powers on the waste management issue, neither he, nor his successor, Martin Cullen, have been able to introduce incineration on a regional basis, as planned. This paralysis is because politicians in Ireland are continually vulnerable to losing their seats, and political careers, due to shifts in public opinion and the vagaries of the PR-STV system. This political instability creates tension which groups like GSE can exploit to further their cause, creating political opportunities for challengers, while restraining government parties. But such instability also encourages politicaias to provide election 'promises' that often cannot be met later.

7.2. BUILD UP TO THE ELECTION: JANUARY 2002

GSE started 2002 by welcoming back international expert Dr. Paul Connett, who addressed a public meeting on 13 January, at the Menlo Park Hotel. Connett's lecture, 'The Incinerator Hasn't Gone Away, You Know', began with Dr. Connett holding up an advertisement from Galway Corporation announcing the achievement of 58% recycling in the city three months after the introduction of the city's recycling scheme. He thanked the Galway audience for achieving such high recycling rates, claiming 'The householders have delivered. This answers the politicians who promoted incineration on the basis that Irish people were too lazy, or unable to recycle'. Connett then asked why five of the seven Regional Waste Plans were awarded to M. C. O'Sullivan, and highlighted the fact that Galway had outperformed the consultants' prediction for 45% recycling by reaching 58%.

Connett was keen to highlight the democratic deficit frame in his comments. He went on to claim that the omission of health risks from the CWP was 'a travesty for Noel Dempsey' adding that 'the level of pollution increases with the level of corruption'. He claimed that Minister Dempsey's changing of the Waste Bill after the CWP's rejection was undemocratic in the aftermath of GSE's campaign. Connett also called on the government to take the lead in promoting recycling through the community, and in attempting to achieve 'zero waste' by 2020. He said promoting incineration was anti-democratic and divisive, while promoting 'zero-waste' could produce harmonious cooperation between the community and the corporate sector. By providing these data on the zero-waste option, Connett and GSE were able to maximise their argument, demonstrating an alternative eco-modern approach was possible through the recycling option.

Local press coverage of the meeting highlighted Connett's call on Minister Dempsey to resign for 'forcing waste incinerators on the Irish people' (*Connacht Sentinel*, 15 January 2002). His accusations about democratic deficit around on the issue were also reported, as was his criticism of the Forfás Report on waste management, which had not mentioned the problem of ash disposal from incinerators. Connett emphasised the idea that the government had acted anti-democratically.

> The government rejected democracy, this is extremely anti-democratic. I don't know how Noel Dempsey can walk across the street in Ireland, because 'he is just tearing up democracy in your face, but its not over .... (Interview, 13 January 2002)

The contestation between the opposing sets of eco-modern waste management options of recycling and incineration was again highlighted at this stage. GSE's campaign continued through the months running up to the election. The group were critical of Minister Dempsey's decision to withhold funding for recycling in Galway, despite the success with recycling up to that point. Minister Dempsey would only fund recycling schemes that had started after January 2002, thus penalising Galway for its early introduction of recycling, and for already surpassing the 2013 EU targets for diversion from landfill. The group interpreted this as a withholding of funding. GSE's criticism of this approach was presented in the following press statement:

> The Minister has shown breathtaking arrogance, just weeks ahead of a General Election, and the people will not forgive or forget this decision. (*City Tribune*, 22 March 2002)

GSE noted that a high recycling rate 'undermines' the rationale behind every one of the seven regional Waste Plans that Minister Dempsey forced through last September, in most cases against the wishes of local councillors' (*ibid.*). The meeting had provided GSE with the opportunity to highlight their eco-modern alternative to the state's own eco-modern policy option of incineration. The Menlo Park Hotel meeting also provided GSE with an opportunity to kickstart their newly politicised campaign. Its main strategic focus of the groups would be to attempt to influence the 2002 General Election in the Galway West constituency. To do so, GSE hoped to rally voters against the largest government party, Fianna Fail. At this point, GSE were moving into their third frame of democratic deficit, and this period would lead to tensions between those who were concerned about the single issue of incineration and its health risks, and those who wanted the campaign to move from the NIMBY and health risks frames into a focus on democratic deficit through the political opportunity of the 2002 general election.

7.3. THE GSE EMERGENCY GENERAL MEETING: EMERGING TENSIONS

In the run-up to the 2002 General Election, GSE would see single issue of incineration give way to an anti-Fianna Fail campaign. Involvement with the formal political process would provide GSE with a degree of success, but it would also result in ultimate disappointment for the overall campaign as the primary focus on the health risks of incineration was now only one part of an attempt to influence the election by highlighting the issue of democratic deficit in the form of a challenge to the Fianna Fail party.

The transition from the first two frames of democratic deficit to the third was not without its problems. Tensions between some members of GSE's committee began to surface at their EGM in March 2002. There were also tensions amongst the groups' activist support, many of whom were prominent members of the local Labour and Fine Gael parties. Suttle had publicly called on all parties to declare themselves against incineration in their election manifestos; something the Green Party had already done. For this reason, Suttle strongly argued at the EGM that GSE should only support the Green's candidate, Niall Ó Brolcháin. This was unacceptable to those who were in some way associated with the Galway West Constituency Branches of Fine Gael or the Labour Party. Labour retained 'heat treatment' as a waste option in their manifesto, and Suttle insisted that this was a form of incineration.

This Labour branch, led by TD Michael D. Higgins, had been committed to environmental issues for many years. In the absence of a strong Green party presence in the past in Galway, Labour had appealed to many environmentalists. Their members were now unhappy at being denied electoral support by GSE. As the GSE campaign moved into its democratic deficit frame, the tactical approach of GSE in establishing their influential alliances with the Green's Ó Brolcháin and the PD's Grealish were alienating other supporters, particularly in Labour, Fine Gael and Sinn Fein. Prominent members of these parties, including their candidates for the forthcoming election, were visible supporters of GSE throughout the campaign. GSE's decision to only back Ó Brolcháin and Grealish would have repercussions for GSE's campaign, particularly in the aftermath of the election campaign.

7.4. MENLO PARK HOTEL DEBATE, 28 APRIL 2002: THE PARTIES SET OUT THEIR POSITION ON INCINERATION

GSE's first attempt to increase access to political structures was now in full swing. The democratic deficit frame was reinforced by GSE's attempt to influence the election campaign and party manifestos. Each party was contacted at the national level and asked to remove any support for incineration in their election manifestos. All the parties except Fianna Fail altered their manifestos correspondingly, indicating the level of influence and political access GSE had attained as a result of their campaign.

However, this tactic was not without its difficulties, as Suttle argued that the Labour party, who had supported GSE more than any other party locally, retained a reference to 'heat treatment' in their manifestos. As Suttle was unable to get this reference to heat treatment removed from Labour's manifesto, she was adamant that GSE could not support Labour's candidate, Michael D. Higgins, who had been to the fore in supporting GSE's campaign. This reaction was not well received amongst Labour's activists in Galway, many of whom had contributed support to GSE during their campaign. Even as GSE were achieving increased access to political structures in their democratic deficit frame, tensions were surfacing and supporters from mainstream parties such as Labour were being left behind by GSE's tactical approach. Labour were also concerned about the single issue nature of GSE's anti-incineration campaign, but continued to oppose incineration, publicly throughout the election campaign, and their statements to date have been consistent on the issue.

In April 2002, GSE organised a debate on incineration among the prospective candidates from the parties. The debate was held under the motion 'That Galway Needs a Waste Incinerator'. Chairing the meeting was David Burke, editor of the *Tuam Herald* newspaper. The format of the meeting was to allow six speakers, three from the government side and three from the opposition, to speak from the platform, while the remaining candidates would speak from the floor. The six invited candidates were Minister Frank Fahey (Fianna Fail), Senator Margaret Cox (Fianna Fail), Minister Bobby Molloy (Progressive Democrats) from the government parties; and Padraic McCormack TD, (Fine Gael) Michael D. Higgins TD, (Labour Party) and Niall Ó Brolcháin (Green Party) from the opposition.

The inclusion on the podium of Ó Brolcháin, a first time candidate, with seasoned TD's indicates the level of support GSE wanted to afford him. Senator Cox had recently voted for incineration in the Seanad, despite her long-standing opposition to it in City Hall. This was to lead to considerable criticism of the Senator during the debate and throughout the election campaign. The meeting began without Junior Minister Bobby Molloy who failed to attend and without Minister Fahey who arrived half an hour late and who left the meeting early.

The first speaker, Labour's Michael D. Higgins, reminded the audience that GSE had provided valuable information and support for the city's councillors on the issue. He said the 'U-turn' on the CWP vote was 'an undemocratic act'. He claimed incineration was a 'voracious, lazy option' and asked why funding had been provided for the CWP but not for an extended recycling programme. Higgins said that Greenpeace opposed incineration for health reasons. He applauded the city's two Labour councillors for their stance on the issue and said that the people of Galway should be able to have a say on incineration. Higgins claimed Labour were 'unequivocally' against incineration.

After his arrival, Minister Fahey was the next to speak. Fahey noted that 'residue waste still had to be dealt with'. The options for this were incineration or landfill. He said that 'all of Europe used incineration', and those who opposed it should 'spell out their alternatives'. He suggested there was a need to look at the latest incineration technology available and said that he had looked and was satisfied with the incineration plants he had seen. He was satisfied that 'incineration was safe and the best way of dealing with waste'. The debate 'should be opened up to the European level, where Green Party Environment Ministers were favouring incineration', according to Minister Fahey.

Following the minister on stage was Oranmore PD councillor Noel Grealish. Councillor Grealish reiterated Galway County Council's opposition to incineration and emphasised his own support for GSE's position. The PDs manifesto did not support incineration, according to Grealish. He said that Ireland was producing 30% more waste per head than any other European country and that there was a need to solve the country's waste crisis. The PDs would create new recycling targets and reduce landfilling. An incinerator would cost £250 million pounds (€317million), while a recycling programme would cost only half of that amount. Furthermore incineration would compete with recycling, which had never been previously used to deal with waste in Ireland. Grealish, in an apparent criticism of Senator Margaret Cox, claimed councillors should either be 'for or against incineration, not abstain in the council vote and vote for it in the Dáil'. Cox and Grealish were expected to contest the final seat in the Galway West constituency, as Grealish had been selected to run at short notice after sitting TD Bobby Molloy resigned his seat. Grealish had supported GSE from the start and was using their campaign to promote his own electoral aspirations. GSE also hoped that the PD candidate would become an 'influential ally' if the coalition was returned to power.

Senator Cox was the next to speak. The senator began by announcing her opposition to incineration. This provoked a reaction from members of the audience, who asked 'why did you vote for it in the Seanad?' The Senator denied supporting incineration in the Seanad, although she voted with the government on the issue. Senator Cox claimed that 'incineration was not part of the plan she voted for'. Senator Cox then said she voted for the plan 'to prevent other counties having a say in our county', a reference to the fact that Galway would have to accept waste from the other counties of Connacht, under the CWP.

At this point, other candidates were given the opportunity to speak from the floor. The first of these was Councillor Lyons of the PDs. He said he made 'no apologies' for adopting the waste plan with incineration. He said while people 'rose to the challenge on recycling, we need full compliance' on waste issues. When asked why he voted for incineration, he said 'he had to at the time' due to the embargo at the Ballinasloe dump. Rosemary 'Dana' Scallon, an Independent candidate and sitting MEP, then presented a populist response. She said that the people of Ireland had rejected incineration. Scallon had opposed incineration due to its health risk, and for the reason that the environmental impact of incineration was 'significant'. Ireland should continue to be a 'green' country, as our organic food industry had achieved a niche in the European market. Scallon also criticised the removal of councillors' powers, saying this was undermining the democratic process. Another of her points was that the EU recognises that large incinerators aren't 'best practice' for waste management, and that there was a need for properly funded recycling schemes throughout the country.

Sinn Fein's Daniel Callanan said that there were health and safety issues as well as democratic issues at stake. He said that the public had backed GSE's petition, and the government had supported this, but that the Minister had closed the debate by taking powers from the councillors. In a reference to the constraints faced by local candidates due to pressure from national parties, he asked that since the PDs had supported the removal of powers in the Dáil, would their candidate Noel Grealish 'now join the Greens?' Although Councillor Lyons had said he had no option but to vote for the waste plan as the rubbish piled up on Galway's streets, Callanan asked why was this rubbish subsequently dealt with without the need for an incinerator? He concluded by saying that 'Galway should not become the capital for waste in Connacht'.

The Green Party candidate, Niall Ó Brolcháin, had been active in establishing the Galway Environmental Alliance, an umbrella group that included GSE. He argued that the Green Party had opposed incineration for

15 years and said they had up to recently been the only party against incineration. The Greens, he claimed, wouldn't go into government with any party that was for incineration. Fine Gael's Padraic McCormack said that people needed to vote to prevent the outgoing government being returned, in order to prevent incineration being introduced in Galway. He said Fine Gael would provide the main opposition to Fianna Fail, and that his opposition to incineration wouldn't be changed if he was elected.

At this point, Michael D. Higgins also claimed that Labour wouldn't go into government with anyone that favoured incineration, reiterating the Labour councillors' stance against incineration. He saw a need for a Waste Management Agency to monitor private waste companies, to ensure accountability. The Galway people 'have shown the government that recycling was viable'. In response, Minister Fahey claimed that the European public's acceptance of incineration had increased with the introduction of better controls. He said that Labour and Fine Gael needed to present their alternative, so they couldn't 'fool the people.' Councillor Grealish refuted the allegation that he wouldn't support GSE's position if elected, and spoke of his participation in GSE's public rally. He said that 'any new government should take incineration off the agenda'.

A member of the audience then asked why Minister Fahey had left the meeting early, after saying he would listen carefully to both sides of the argument. A discussion then developed about the destination of recycled plastics from Ireland, which were said to go to a plant in Slough in England. The meeting ended with a call for scientific evidence to back up Minister Fahey's claims about the safety of incineration. Fears were expressed about the results of dioxins entering the water supply, and incinerations potential to increase global warming. Independent candidate, Eileen Manning, accused Minister Fahey of 'running away' from the meeting.

Press reports picked up on the widespread opposition to incineration amongst Galway's candidates, as well as Minister Fahey's argument about the safety of incineration. They also reported on some candidate's concerns about the undermining of the democratic process in the handling of the CWP. GSE were reported as claiming that 'there was no support for incineration, outside of the cabinet table' (*Galway Advertiser*, 2 May 2002). GSE were extending their democratic deficit frame, targeting the core of power in Irish politics, the cabinet. The public debate and changes in the party manifestos represented the clearest indication that GSE's campaign had moved beyond the single issue of incineration, through the health risks and NIMBYism frames, and was now fully in the democratic deficit frame, targeting the coalition cabinet, and Fianna Fail in particular. As GSE moved squarely into their democratic deficit frame, their choice of such 'influential allies' would have a major impact on their campaign. However, their support for the candidates from the PDs and Greens only would prove to be contentious and would have implications for GSE's campaign.

Throughout May 2002, the general election campaign gained momentum. Waste management and incineration became an issue for all parties in the election, something that reflected the extent of GSE's framing of the issue, with all parties except Fianna Fail declaring their opposition to incineration following the Menlo Park Hotel debate. At this point, GSE actively called on the Galway voting public to vote strategically against Fianna Fail, on account of their support for an incinerator in the city. The fact that GSE were able to influence all parties except Fianna Fail to alter the election manifestos in regard to incineration demonstrates the influence they had at that stage. While a certain amount of discord had developed over GSE's initial backing of the Green's Niall Ó Brolcháin, GSE's inclusion of the junior coalition partner, the Progressive Democrats, amongst the parties they urged support for, was not welcomed by the opposition parties and would backfire in the long term. While GSE hoped to have 'influential allies' in the coalition parties as well as the opposition, the demands of the Irish political system would ultimately supersede any local alliances created by GSE.

7.5. THE PARTY'S ELECTION MANIFESTOS: REJECTION OF INCINERATION

This section examines the parties' manifestos in the light of GSE's intervention. GSE's campaign had at this point influenced all parties except Fianna Fail. This opposition to oppose incineration for populist reasons was spelled out in the parties' election manifestos. Fine Gael's waste policy 'Protecting Our Future' incorporated a system of exhaustive reuse, tax incentives for recycling, and reductions in packaging levels. The party favoured a 'zero waste' strategy to reduce landfill levels and opposed incineration on health grounds. Fine Gael expressed the need for a waste policy that could match economic growth rates and saw an inevitable link between both. Their policy statement also incorporated the appointment of a waste commissioner, as well as a special Oireachtas committee that would have six months to draw up an alternative waste strategy. Fine Gael also included the core principle in their waste policy that called on business and industry to incorporate design methods that facilitated waste reduction and included regulation and incentives to support that initiative.

Major reuse, recycling and composting projects would be adequately funded, with a polluter pays approach to service charges being included in their strategy.

Fine Gael were concerned that Fianna Fail could return three seats in Galway West and that this would ensure an incinerator would be built in the city. McCormack claimed that if Fianna Fail got the three seats, 'the people would have to accept this as a democratic vote' in favour of incineration (*Galway Advertiser*, 16 May 2002). Fianna Fail was also aware of the populist nature of Irish politics, having exploited this populism as part of their 'catch all' approach to elections over the years.

The Labour Party had been active in GSE's campaign from its earliest stages, with local TD Michael D. Higgins providing strategic advice for the group since its inception (M. D. Higgins, interview, 20 November, 2000). The two Labour councillors, particularly Alderman Connolly, had taken a strong stance against incineration. Labour's policy statement, *Waste Matters*, called for a new waste policy, based on reduction, reuse and recycling, which would be properly funded. Labour were opposed to incineration, and proposed investment in a recycling industry with reliable markets for recycled products. Waste collection services would be structured to facilitate this process. Labour would establish a National Waste Management Agency to facilitate the new waste strategy, and link it with the National Development Plan. Labour proposed to divert more funds from the NDP to build a recycling infrastructure in the country and use funds for public education projects on the issue. Waste disposal that relied on landfill would be minimised under Labour's waste strategy.

Labour's plans for a National Waste Management Agency would include the funding of local authorities to clean up existing waste facilities such as landfills, provide separation and recovery infrastructure for household, industrial and agricultural waste, provide new household and hazardous waste facilities and develop markets for recycling. The 'polluter pays' principle arrangement was to be introduced for collection services. Labour vowed to ensure:

> that the EPA has the full technical expertise and financial resources necessary to ensure its integrity and independence and to provide for the best and most efficient regulatory and control system internationally for the waste sector. Labour will also ensure that the EPA will be provided with the necessary powers to stop immediately any activism or intended activities where it believes a risk to the environment or human health may arise. (*Waste Matters:* Labour's alternative strategy for waste management, February 2002, p. 4)

The Green Party candidate, Niall Ó Brolcháin launched his campaign by claiming that the general election would let the voters decide on the incineration issue. He called on the local authorities to use their powers to amend any adopted waste plan and to omit incineration altogether. He claimed the waste crisis had 'been allowed to develop by the parties in power' (*Connacht Sentinel*, 6 February 2001). Throughout the election campaign, it was clear that GSE had made available all of their scientific evidence about dioxins to Ó Brolcháin. Indeed, GSE press releases were often reported alongside similar statements from Ó Brolcháin during the election campaign.

The Green Party's waste policy included a plan for dealing with the waste crisis without incineration. The Green's view of incineration was that

> There are sufficient grounds to believe it is an unsafe technology. It allows householders and companies to produce high levels of waste without taking full responsibility for it. ie. It allows the producer pays principle to be ignored. It discourages recycling as a guaranteed level of waste is required for the lifespan of the incinerator. It is an extremely expensive technology, both in financial and environmental terms. (Green Party Election Manifesto, April 2002)

7.6. ELECTION OUTCOME

As late as mid-April, the Progressive Democrats had changed their waste policy to oppose incineration, leaving Fianna Fail as the only remaining party supporting incineration. The PDs revived election manifesto called for waste reduction, recycling less use of landfill and no mass-burn incineration. The manifesto contained the following section on incineration:

> The Progressive Democrats are confident that Ireland's problems can be solved without resort to mass-burn incineration. Mass burn incineration is essentially a twentieth century technology, a generation of technology which we in Ireland have never embraced. Essentially, mass-burn involves the burning of untreated and unsorted waste, and burning it gives rise to serious problems: Conflict with Recycling: local authorities have to enter long-term agreements to supply fixed quantities of waste to incinerator operators. Otherwise, they face financial penalties.
>
> This is in direct conflict with any long-term strategy to cut waste volumes by recycling and reduction. Traffic: Big incinerators generate big volumes of traffic. Emissions: The EU has set tight limits on toxic emissions from incinerators but they are only effective with tight enforcement. In 1999, and 2000 the eleven municipal incinerators in the UK breached their limits more than fifty five times. Residual Ash: Household rubbish yields highly toxic product when burnt. Modern plants filter the toxins but dioxins remain in chimney ash. (Progressive Democrats Election Manifesto, 2002, p. 51)

The PDs used the incineration issue to target Senator Margaret Cox of Fianna Fail who had opposed incineration initially, but then voted for it in the Seanad. Grealish had been critical about Cox's Seanad vote during the election campaign, as the PDs had to try to prevent a Fianna Fail candidate taking Molloy's PD seat in Galway West. Their candidate Noel Grealish had also been the most vocal critic of Senator Cox's position on the issue during GSE's pre-election meeting on 28 April. When contacted, Grealish's office asserted that 'he had played a major part' in influencing the PDs rejection of incineration in their manifesto (Grealish Constituency Office aide, in telephone conversation, 12 May 2004).

By early May, it was apparent that the Green's Niall Ó Brolcháin was unlikely to join the two main opposition candidates who supported GSE, Labour's Michael D. Higgins and Fine Gael's Padraic McCormack in the Dáil. Grealish had been given a seat at the platform at GSE's Menlo Park meeting, rather than Fine Gael's Michael McDonagh, who had also been a long-time supporter of the anti-incineration stance.

The results of the vote in the Galway West constituency, for the 2002 General Election were as follows:

Fianna Fail 41.3%, Fine Gael 16.9%, the Progressive Democrats 12.5%, Labour 10.5%, Sinn Fein 5.6%, the Green Party 4.4% and others taking 8.7%. Fianna Fail held their two seats, those of the Minister Eamon Ó Cuív and Frank Fahey (*Connacht Sentinel*, 15 May 2002). Fine Gael held the seat occupied by recycling champion Padraic McCormack. The Labour Party's Michael D. Higgins also held his seat, while Noel Grealish achieved 'the impossible coup' (*Connacht Sentinel*, 15 May 2002) by taking the final seat. An analysis of the transfers for Grealish indicates that he was able to draw transfers from all of the candidates who supported GSE. The breakdown of these transfers was as follows:

Independent candidate Dana Scanlon, who had opposed the incinerator, provided Grealish with 112 transfers, while the Green's Niall Ó Brolcháin provided 332. The importance of Grealish's ability to attract transfers from a variety of sources was picked up on in the media reports on the election, which hailed Grealish's result as one that 'defied all expectations'. GSE's electoral support for Noel Grealish was acknowledged by Grealish's constituency office when I contacted them, and was reported in the press (*City Tribune*, 14 May 2002). Independent candidate Dana Scallon, who had opposed the incinerator, provided Grealish with 112 transfers, while the Green's Niall O Brolchain provided 332. The importance of Grealish's ability to attract transfers from various sources was picked up on in the media reports on the election, which hailed Grealish's result as one that 'defied all expectations' (*City Tribune*, 14 May 2002).

Grealish had received significant support form GSE, alongside the Green's Niall O Brolchain. His Turloughmore ward ran alongside GSE Conchur O Bradaigh's locality in Castlegar, which was one of the proposed sites for the incinerator. The area had been a source of support for GSE and had been notably targeted in GSE's trademark 'skull and crossbones' placard campaign proclaiming 'no to incineration' and 'dioxins kill'. This placard campaign had been led by Conchur O Bradaigh himself.

Fianna Fail were despondent at not achieving a third seat in Galway West. A 'Prime Time' opinion poll, broadcast by RTÉ before the election, had forecast 51% of first preferences for fianna Fail and 3 seats for the party in Galway. Fianna Fail were predicted to increase their vote nationally. However, that opinion poll had come out before the Menlo Park Hotel waste debate. Subsequently, Grealish was able to benefit from transfers from anti-incinerator candidates such as Higgins, McCormack and Ó Brolcháin, as well as benefiting greatly from the support of his two city-based PD running mates, Donal Lyons, and Declan McDonnell.

On the 15th count, the Greens' Ó Brolcháin transferred 322 votes to Grealish, a remarkable amount from candidates and from parties which had so little in common. Grealish had been singled out for overt support by GSE along with the Green's Niall Ó Brolcháin. His Turloughmore ward ran alongside GSE's Conchúr Ó Brádaigh's locality in Castlegar, which was one of the proposed sites for the incinerator. The area had been a source of support for GSE, and had been notably targeted in GSE's trademark 'skull and crossbones' placard campaign proclaiming 'no to incineration' and 'dioxins kill.' This placard campaign had been led by Conchúr Ó Brádaigh, himself. The final contest between Grealish and Margaret Cox from Fianna Fail was also determined in part due to their position on the incineration issue. Grealish received 332 transfers from Ó Brolcháin, despite being a country candidate who ran a short campaign, while Cox, from Ó Brolcháin's own ward, only received 312 transfers from the Green candidate. Michael D. Higgin's surplus also favoured Grealish, as he picked up 531 transfers to Cox's 321. Grealish had received a remarkable amount of transfers from city-based opposition candidates, more than the PD's two city-based candidates had managed. Local press reports confirmed the fact that the incineration issue had been a factor in Grealish's success with Fianna Fail concerned about candidates who could 'get an entree to Green votes' while acknowledging the significance of Grealish's support for GSE:

> Noel Grealish had attended anti-incinerator meetings and has spoken against the incinerator concept. (*City Tribune*, 21 May 2002)

It is interesting to note that as a rural candidate with no experience of Galway city politics, Noel Grealish received two-thirds of his total vote from transfers from the city. Grealish had only decided to stand weeks previously in light of Molloy's unexpected decision to stand down. Minister Molloy had been critical of GSE, accusing them of 'scaremongering' and using unreliable information. In contrast, Grealish had opposed incineration while serving on Galway County Council and appeared on GSE's platform on a number of occasions.

Further evidence of transfer patterns from 'green' voters to Grealish can be seen in the analysis of the constituencies where transfers affected the overall outcome in the 2002 election. Overall, transfers played a part in a seat being won in 14 different constituencies, including Dublin, Cork, Limerick as well as Galway West (Gallagher, March, & Mitchell, 2003, p. 107). Gallagher's study highlights the Green Party transfers that went to Noel Grealish as a significant factor in Grealish's success at the expense of Margaret Cox (*ibid.*). As Grealish was a rural candidate only known in the city because of his support for GSE during the run up to the election, it can be said that Grealish's support for GSE tipped the result in his favour. Of course, the PD's ability to maximise vote transfers among its candidates must be noted as a factor in their success.

The incineration debate also became a political issue in the 2004 Local and European elections. In the aftermath of decision to allow incineration in Meath and Cork, the campaign to oppose incineration has resumed. No decision is expected on an incinerator in Connacht until after the 2004 elections, as the government attempt to avoid the issue impacting upon their candidates. In a recent interview, Environment Minister Martin Cullen criticised groups that had opposed incineration. According to Mr. Cullen, these groups should 'grow up, be mature, and stop all the nonsense' (*Irish Times*, 23 January 2004). The incineration debate continued into 2005, but at this point, no incinerator has been built in Galway, or elsewhere in Ireland.

7.7. GSE SINCE THE GENERAL ELECTION: POLITICAL CLOSURE AND RESTRICTIONS OF ACCESS TO POLITICAL OPPORTUNITY STRUCTURE

The campaign against incineration had run for over two years by the summer of 2002. It had witnessed successes and setbacks, and a degree of campaign fatigue had set in after the intensity of the General Election campaign. The return of the Fianna Fail-Progressive Democrats government was a concern for GSE, and any advantage gained from supporting the PDs candidate began to fade as the new government continued to press its incineration-based waste policy. The new Fianna Fail Environment Minister Martin Cullen appeared to be committed to incineration.

The final months of 2002 saw GSE's campaign in a period of crisis. Conchúr Ó Brádaigh left to establish an engineering business, but remained available as an 'advisor' to the group. Many supporters of the group had felt uneasy about the support shown for the PDs at the election. The treatment of sympathetic parties especially Labour and Fine Gael over the stances on incineration had taken its toll on GSE's activist base. GSE had also become closely linked with the Green Party candidate, Niall Ó Brolcháin, and shared somewhat in his electoral defeat.

GSE's sense of crisis was highlighted in a letter to the local press sent by long-standing GSE activist, Brendan Smith, who had organised GSE's successful petition in April 2000. The full-page letter ran under the headline 'Was Waste the PDs Sacrificial Lamb at the Altar of Power?' (*Galway Advertiser*, 5 December 2002) The letter began by reminding the public about the PDs pre-election 'strong anti-incineration policy that represented a major U-turn for the party' (*ibid.*). According to Smith, it was this policy change that helped the PDs to electoral success in Galway:

> This seismic policy shift had a dramatic effect on the election results in Galway West and 'Lazarus-like' the PDs merrily walked away from what many renowned pundits were saying was to be their political funeral. (*ibid.*)

Smith's letter went on to highlight GSE's campaign to have all parties publish their waste policies before the election. It also highlighted Fianna Fail's 5% drop in Galway West that went against the national trend. Smith attributed that drop to GSE's campaign, and he highlighted the fact that Galway West returned three candidates who professed to be against incineration. Smith refers to Fianna Fail's waste policy as 'a grave error of judgement' (*ibid.*).

This is contrasted with what is referred to as the PD's 'Road to Damascus conversion' on the waste issue (*ibid.*). The PD's victory is attributed to their 'committed campaigning against incineration' locally as well as 'his high transfers from normally diametrically opposed sources such as the Green Party' (*ibid.*). However, the action of both government parties since the election has 'left a bitter aftertaste that places a serious question mark over the founding principles of the PDs (*ibid.*), according to Smith. He accused the PDs of betrayal on the incineration issue and called into question that party's stance of promoting honesty and transparency in Irish politics.

The following questions were then posed by Smith

1. 'Why have the PDs reneged on their promise to stop municipal incineration?'
2. 'How can the PDs stay in government if such a cornerstone of their party policy is not being implemented?'
3. 'Why is Galway City being denied full funding under the Waste Management Infrastructural Grant Scheme for its recycling on the ridiculous grounds that we needed to start before January 1st?'
4. 'Why is the Galway city recycling household waste collection system not being promoted by the government as the way forward for other local authorities?'
5. 'Where is the health impact study on incineration that was promised by the government before the last election?'
6. 'Why are the concerns of racehorse breeders such as Aiden O'Brien being ignored when they say that Ireland's bloodstock breeding and horse training industry will be destroyed by the construction of dioxins producing incinerators in what is Europe's most dioxin free environment?'
7. 'Why are government departments and their consultants allowed to promote Belgium and Denmark as models of good waste practice when the former has one of the worst dioxin environments and the latter one of the worst recycling rates in Europe?' Because both countries serve as world leaders in manufacturing incinerators and both are gearing up to build the Irish incinerators:
8. Why are government departments and their consultants allowed to promote incinerators as solutions to waste reductions when they actually create 33 tons of toxic ash laced with dioxin for every 100 tons of refuse incinerated?'
9. 'Does the government not realise the carcinogenic dioxin emission from incinerators will undermine the health rates of our people and lead to higher cancer rates?'
10. 'Why was Irish Glass, a company of national strategic importance to the country's recycling efforts, allowed to close down?' (*Galway Advertiser*, 5 December 2002)

The Progressive Democrats continued to be criticised by GSE throughout the spring of 2003. However, GSE's need for new activists and a fresh direction became apparent with the news that Aine Suttle was going to step down due to health reasons. With GSE's two leading activists withdrawing, the mantle of leadership was passed in the main to those who had been active in the election campaign of Niall Ó Brolcháin of the Green Party.

GSE called an AGM for 20 March 2003, at the Atlanta Hotel in Galway. However, this meeting coincided with the outbreak of war in Iraq, and many politically minded people stayed at home to follow events on evening news bulletins. Subsequently the meeting drew an attendance of only 12 people. The meeting began in some confusion, as GSE's rules stated that 20 people had to be present to elect a new committee. Eventually enough late-comers arrived to allow the meeting to commence. The Green's Kieran Cunnane became the chairperson of the group, with Chris McCarthy becoming treasurer. Aine Suttle announced her resignation from the committee.

At the meeting, GSE's tactic of backing the PDs at the 2002 General Election was questioned. One speaker criticised the strategy, asking 'don't you realise that you've been sold a pup', in relation to the PD's subsequent agreement of a Programme for Government that included incineration. The meeting concluded with a decision to continue public criticisms the PDs. GSE released a press statement that called on the public to picket the PDs party conference, which was taking place on 11 April at the Corrib Great Southern Hotel. GSE accused the PDs of 'hoodwinking' Galway's voters on the incineration issue, referring to that party as 'Fianna Fail's lapdog' (*City Tribune*, 1 April 2003). It also called on the PDs to carry through their election promises and reject incineration:

> Michael McDowell asked us not to trust Fianna Fail, on their own. Mary Harney said the PDs would be the watchdog over Fianna Fail, instead they are turning out to be their lapdog. It's now looking like a vote for the PDs was a vote for incineration after all. (*ibid.*)

While GSE's campaign did go into a period of decline throughout 2003, the granting of planning permission for incinerators in Meath and Cork revived interest in incineration as an election issue. The first month of the year saw GSE's Kieran Cunnane release a press statement claiming that 'the signs are ominous for an incinerator in Galway' (*City Tribune*, 23 January 2004), in the wake of An Bord Pleanála's decision to follow government policy on the incineration issue. Cunnane was critical of this position, and warned about the consequences of the promised National Infrastructure Board, adding that there was 'an axe hanging over An Bord Pleanála's head' (*ibid.*).

Cunnane reiterated GSE's switch from an anti-government to an 'anti-Fianna Fail' position at the 2002 General Election, claiming that 'the GSE tactic paid dividends with Margaret Cox losing out narrowly to the PDs' (*ibid.*). He also expressed GSE's disappointment with the PD's performance since the election and said GSE would change their future electoral strategy to an anti PD, as well as anti Fianna Fail position, saying 'talk is cheap, but standing up for your stated pre-election principles and honouring your constituents' wishes in the Dáil are what really counts'. In direct reference to the constraints faced by the PD's since the election, Fidelma Healy Eames, who was standing in Grealish's Oranmore district, suggested that

> Despite his ongoing promises to keep incineration out of Galway, we know when the Protection of the Environment Bill was going through the Dáil, Deputy Grealish and his Fianna Fail and PD colleagues supported the proposal to take power away from councillors in relation to the adoption of waste management plans and give it to the Managers. (*Galway Independent*, 4 February 2002)

These criticisms were echoed by the Green Party's Niall Ó Brolcháin, who had run for election to the city's South Ward in the 2004 local elections. Ó Brolcháin called on the PDs to clarify his position on incineration, pointing out how they were now trying to convince Galway constituents that incineration wouldn't be used for waste management in the West.

> This is a marked change from their stance in the General Election when they claimed to be steadfastly opposed to mass-burn incineration. The reality is that the government is planning to build incinerators immediately in Dublin, Cork and Meath ... and ultimately Galway will not be spared. (*Galway Independent*, 4 February 2004)

By this time, the central figures in GSE's campaign, Conchúr Ó Brádaigh and Aine Suttle had disengaged from the group. This loss of the campaign's two most effective mobilisers of the key resources of expertise and networks would have a significant impact on GSE, which was ultimately co-opted by the local Green Party. In a recent Community Forum newsletter, one of GSE's new spokespersons, Kieran Cunnane, discussed the public consultations which were held on the new Connacht Waste Plan, which was due for adoption in the spring of 2005.Cunnane was the Green Party candidate in the 2004 local elections, but was unsuccessful. He states that he complained at the meeting about what he claimed was 'propaganda' about incineration in the council's information booklet on incineration. He also reiterated GSE's concerns about the health risks and costs of incineration. The article includes a criticism of the PD's call for a central incinerator for the whole country, his first public statement on incineration since supporting GSE's campaign in the run up to the 2002 general election. Cunnane's views on Grealish and the PD's points to some continuity in their perspective of what had happened during their attempt to influence the 2002 election campaign:

> The background is that prior to the 2002 General Election, GSE attacked Fianna Fail only. We refrained from attacking the PDs because they passed a motion (which Noel had a big input into), precluding mass burn incineration. He has made his and the PDs

position finally very clear (we always suspected this); they are for incineration, but not in my constituency. Galway for a Safe Environment are not about being NIMBY's and are not in favour of fobbing off an incinerator to some other unfortunate community, we are opposed to it anywhere in the country. (Cunnane, p. 6, *Community Forum Newsletter*, April 2005)

7.8. CONCLUSION

This chapter has looked at the shifting political opportunity structure surrounding the 2002 General Election. A series of events have been examined to illustrate the shifts in the dispute as the incineration issue became a part of the electoral campaign. These critical events included the January 2002 Menlo Park Hotel meeting of GSE, The GSE EGM in February where GSE decided to make incineration an issue in the General Election, the debate organised by GSE in April 2002, the General Election campaign throughout April and May 2002, and the 2002 General Election outcome, returning the pro-incineration parties to government.

The main strategy of GSE at this time was their attempt to influence the political parties to change their election manifestos to include a rejection of incineration. Unfortunately for GSE, election manifesto promises in Irish politics are often lost in the negotiation to form post-election governments. It would seem that the PDs adopted an anti-incinerator stance to initially facilitate their candidates in Galway and elsewhere, while preparing to later drop the policy to facilitate a coalition with Fianna Fail. In the Seanad Senator Cox suffered some criticism for her support for a waste plan that included incineration despite their initial public support for GSE's position. This U-turn may have been a factor in Cox's subsequent electoral defeat at the hands of the PD's own candidate, who used GSE's election debate to attack Cox's change of position to his own advantage.

The incineration debate again became an issue in the political arena, in the 2004 local and European elections. In the aftermath of decisions to allow incineration in Meath and Cork, the campaign to oppose incineration has resumed. However, the lack of any orchestrated public protests surrounding the new Connacht Waste Plan in 2005 indicates both the extent of the co-option of GSE by the local Greens, and the attempts by the state to incorporate local representation into a consultative process in a way which avoided the creation of 'increased access' to political opportunities which had occurred at the beginning of GSE's campaign. As GSE's campaign focused on the third frame of democratic deficit, increased access to political structures was achieved through the political opportunity of the 2002 election campaign. However, GSE's choice of the PD's candidate as an influential ally would damage their campaign, as supporters of opposition parties were left behind by this strategic choice. As Fianna Fail was returned to power, GSE's campaign went into a period of decline, while the PD's were unable to continue his support for GSE due to political pressure from the coalition cabinet, indicating the constraints placed on local politicians as pre-election alliances give way to post-election negotiation for government.

CHAPTER 8

CONCLUSION OF THE STUDY OF THE GSE CASE

8.1. SIGNIFICANCE OF THE GALWAY CASE STUDY

I have presented an 'issue history' (Szasz, 1994, p. 162) case study of GSE's campaign against incineration. This approach has utilized different theories to examine GSE's campaign. It has outlined the study in the context of a NIMBYist campaign that utilized three main frames of health risks, NIMBYism and democratic deficit. The campaign was set in the second of two phases of environmental campaigning in Ireland. The first phase established community opposition to multinationals, whereas the second phase witnessed NIMBYist campaigns against infrastructural projects, such as incinerators, which the state tried to introduce to deal with the post-boom waste crisis.

Within the context of this second phase of campaigning, I have considered the political opportunities that emerged for the environmental movements that wished to challenge the state's waste management policy. These 'bottom-up' challenges to the state involved campaigns against the state's eco-modern waste solutions, which promoted some recycling in addition to regional incinerators. The state's waste management agenda was developed after the formation of the EPA in 1992 and attempted to implement a waste framework that was built through the 'top-down' directives of the EU.

GSE's campaign emerged in the aftermath of first phase challenges to multinational pollution. First phase campaigns in the pre-boom era challenged the siting of multinational polluters. These first phase challenges created a basis for local mistrust of multinationals, as well as a mistrust of the state, whose policies promoted multinational development. The suspicions of local communities could be exploited by second phase groups like GSE, who framed its campaign around issues such as health risks. This exploitation of mistrust was facilitated by the state's reliance on multinational-led solutions, such as incineration to the waste crisis.

The study has addressed the GSE case through three main theoretical positions. These are Tarrow's four salient changes to POS, Szasz's transition from NIMBYism to ecopopulism, and Snow and Benford's understanding of framing issues. These three theoretical positions are utilized to provide clarity around the significance of GSE's campaign. The first of these theories is Tarrow's four most salient changes to POS. These four changes

(i) influential allies
(ii) divided elites
(iii) unstable alignments
(iv) increased access

can be applied to key moments in GSE's campaign to demonstrate where access to the political structures of the state are gained or constrained although Tarrow's four changes can only go so far in explaining these key moments. This study applies the four salient changes to key moments of GSE's campaign to demonstrate the changes in the political opportunity structures surrounding these issues follows:

8.1.1. Sidney Tarrow's Theory

(i) *Influential Allies*

GSE forged many alliances throughout their campaign. Their committee was an alliance of experts and experienced campaigners, who facilitated the mobilisation process. Its alliance with Paul Connett provided the 'interest-led' science that underscored their challenge to the state's waste policy. Political alliances provided further support, while alliances with other anti-landfill and anti-incinerator groups allowed GSE to portray its campaign as more than NIMBYist. One of GSE's 'influential allies', the PD's, provided a 'perceived' (Tarrow, 1988, p. 77) opportunity to have an ally within the coalition parties' ranks. However, this perception of opportunity through alliance proved to have unintended consequences, as the demands of government superceded the PD's promise of support for GSE after the 2002 election.

(ii) *Divided Elites*

GSE attempted to exploit the divisions that occurred between the state and the Councillors elected to the city and county councils. It had some success with this policy, particularly when they influenced the city council's rejection of the CWP. GSE also had some success in exploiting the state's reliance on technocratic advisors and consultants. This division provided GSE with an opportunity to present its 'interest-led' arguments against incineration and allowed GSE to portray the state as somewhat secretive, opening the democratic deficit frame. GSE had less success exploiting the potential divisions between the state and the EU. This was due to the state's adherence to the ecomodern solutions of incineration and recycling. GSE was only able to challenge the state about its emphasis on incineration over recycling, while both sides rejected the EU's least-favoured option of recycling.

(iii) *Unstable Alignments*

Tarrow's theory of unstable alignments doesn't account for the demands of coalitionism in the Irish political system. While GSE attempted to develop its campaign from a single issue, anti-incineration campaign to a political challenge against Fianna Fail during the 2002 general election any division between the coalition parties of government, the PDs and Fianna Fail, were overcome due to the demands of forming a new government after the election. While GSE had allies within all the parties except Fianna Fail, its attempt to exploit the instability of the coalition parties, alignment failed, and its movement declined as a result.

(iv) *Increased Access*

Most of the key attempts at leverage mentioned above were undertaken to increase access to the political structures of the state. The political opportunities that GSE attempted to exploit emerged at certain moments of their campaign. Some provided temporary access, while other 'perceived' opportunities proved to be more illusory. The first moment of increased access came from the state's attempt to introduce incineration without mentioning the potential health risks it posed. GSE used this opportunity to frame and mobilize their campaign while portraying the state as somewhat dishonest in its approach to the issue. The second opportunity for increased access stemmed from the council's vote on and rejection of the CWP.

The third and most significant moment of increased access emerged from the opportunity of the 2002 general election. GSE was able to exploit the populist concerns of the parties and their candidates, who feared an electoral backlash on the incineration issue. GSE were able to increase their access to the parties in a significant way, by getting all parties except Fianna Fail to change their manifestos and reject incineration. This represented a major success for GSE, as a NIMBYist group, who were able to influence politics on a national level by having the manifestos altered.

However, even this increase in access proved to be temporary, as the coalition government of Fianna Fail and the PDs were returned to office. This reversed GSE's increased access, as the new programme for government again included incineration as a waste management option. Furthermore, GSE's attempt to increase access through its alliance with the PDs affected its activist support, many of whom were members of the opposition parties. GSE's attempt to increase access through the opportunity of the 2002 general election ultimately led to a loss of access, and a loss of support for the campaign.

GSE's case demonstrates the interconnectedness of Tarrow's four salient changes, both in improving access to the political structures of the state, but also in the loss of not only the access and opportunities gained, but of movement momentum itself, as co-option from the disarray of such setbacks occurs. This was represented in GSE's case by its co-option by the local branch of the Green Party, who have not been as vocal on the incineration debate as the single issue GSE campaign was. This co-option of a branch of civil society by a formal political party represented an outcome not foreseen in Tarrow's theory.

8.1.2. Andrew Szasz's Ecopoulism

According to Szasz, (1994, p. 69) movements can develop from 'NIMBYism into radical ecopopulism'. By using Szasz's issue history approach to this case study, I have demonstrated the failure of a NIMBYist group like GSE to emerge as radical ecopopulist. This progression failed to occur for a number of reasons. The first of these is the small number of key personnel involved with GSE's committee. These committee members held various political perspectives, some more radical than others. The only political point that unified this group was their opposition to incineration. So, at the moment when GSE attempted to politicise its campaign during the 2002 general election, this lack of a clear political focus lead to its misinterpretation of the 'perceived' opportunity of political alliances.

The two senior figures in GSE, Suttle and Ó Brádaigh, shared little by way of political ideology outside of their opposition to incinerators. Suttle was an eco-feminist campaigner, while Ó Brádaigh was a radical republican who was opposed to the state. While Suttle's alliance with Paul Connett provided GSE with 'interest-led' data for its campaign, he was also a single issue anti-incineration campaigner whose own political ideology failed to go beyond that issue. GSE's political alliances failed to deliver any radical political transformation for their campaign. While Sinn Fein supported GSE, Ó Brádaigh's involvement in republican splinter group politics precluded any further alliances with that party. The Greens had yet to make a breakthrough in Galway or national politics, although the Galway Safe Waste Alliance's Niall Ó Brolcháin has since gained a city council seat, in the 2004 local elections. However, by that stage GSE had been co-opted by the Greens, and the incineration issue became just another part of the Green's overall environmental manifesto. While Fine Gael had offered GSE support before the election, they seemed to lose interest afterwards, in the same way that the PD's did. Perhaps it was the Labour Party that offered GSE the best option for political transformation.

Labour TD, Michael D. Higgins, had offered GSE support and direction from the early part of their campaign. Labour Councillor Catherine Connolly, a neighbour of Aine Suttle, and strong supporter of GSE in the city council, became mayor in 2004, where she continued to oppose incineration. Connolly was central to the omission of incineration from the City Development Plan during her tenure. However, GSE never followed through on a possible alliance with Labour at local or national level, due in part to Labour's neglecting to remove a reference to 'heat treatment' in its manifesto in 2002. For Suttle in particular, the single issue of incineration superceded the political opportunities posed by an alliance with Labour. When GSE supported the Greens and PDs during the 2002 election, the numerous members of the local Labour party, who were historically active on environmental issues in Galway, then lost interest in GSE's campaign. This subsequent loss of support was compounded by the departures of both Suttle and Ó Brádaigh. Reliance on an elite committee constrained by their focus on a single issue, combined with a degree of political misjudgement ultimately proved costly for GSE.

8.1.3. Snow and Benford's Framing Theory

This study has utilized Snow and Benford's theory of 'Master Frames and Cycles of Protest' to illustrate the significance of GSE's framing of the significant issues at key moments of its campaign. Although the study highlights three main frames, these frames are somewhat contradictory and overlapping. As GSE's campaign unfolded, these three frames came to represent key stages of resource mobilisation and attempts to gain access to

POS. The frames also represent an attempt to move beyond NIMBYism into a more political campaign. The three frames are as follows:

(i) *Health Risks Frame*

GSE's campaign emerged from the political opportunity of the state's omission of health risk in the CWP. Through its alliance with Connett and their utilization of internet technologies, GSE was able to frame its campaign around the health risks issue. The health risks frame was facilitated by the concerns about toxicity created by first phase campaigners against multinationals. It was also assisted by the state's own duplicity on the health risks issue and inability to allay the fears of the public and local councillors on the issue. The health risks frame provided GSE with increased access to local politicians, and led to the rejection of the CWP.

(ii) *NIMBYism Frame*

GSE was concerned about being portrayed as NIMBYists. To avoid this, they attempted to forge networks with other anti-incinerator campaigns in Cork and Meath. GSE also forged networks with local anti-superdump campaigners, to prevent the state from playing both campaigns off each other. In addition, GSE exploited the opportunity of the 2002 election to politicise their campaign at a national level. In so doing, GSE hoped to distance itself from NIMBYism, and gain access to the formal political structures and parties of the state. However, one of the contradictions of GSE's campaign was the overlapping nature of the NIMBYism and health risks frames. GSE exploited public concerns about health risks, together with political concerns about an electoral backlash on this issue. Local politics also had a NIMBYist aspect, which GSE could exploit. The persistence of the health risks frame throughout GSE's campaign impinged on the other two frames, and ultimately constrained GSE's political campaign.

(iii) *Democratic Deficit Frame*

GSE attempted to exploit the state's duplicity on health risks and their 'confrontative' response to GSE's campaign as part of their democratic deficit frame. GSE was able to build upon the mistrust of politics created by first phase campaigners, as well as benefiting from a general malaise among the Irish public regarding post-tribunal politics in general. GSE was able to exploit the state's reliance on technocratic advisors, who created a culture of secrecy in the state's various departments. GSE attempted to extend its democratic deficit frame by attempting to influence the 2002 general election. Again the health risk and NIMBYist frames intervened, as GSE misinterpreted the political landscape. Its single issue health risks campaign ended up in the choppy waters of electoral politics. My study indicates that while Snow and Benford's framing theory can go some way towards providing an understanding of how campaigns frame issues, these frames are never distinguished in GSE's case, as each frame overlapped and influenced the other.

Using Snow and Benford's (1992, p. 136) discussion of how meanings are 'amplified', old meanings are 'transformed' and new meanings are 'generated', my study has charted the emergence of GSE's collective action frames. This first frame, health risks, was 'amplified' (*ibid.*) through GSE's public and media campaigns. This led to the mobilization tactic of the public petition, and also influenced the councilors' rejection of incineration as part of the CWP. GSE's second frame, NIMBYism, was 'transformed' (*ibid.*) by its attempts to create networks with other NIMBYist groups. This may be seen as the 'amplification' of local NIMBYist interests into a national NIMBYist movement. However, we can say that GSE's attempts to influence political structures went beyond the established understandings or 'old meanings' of a NIMBYist campaign. Finally, GSE's campaign generated 'new meanings' (*ibid.*) around the anti-incinerator debate, as it introduced concerns about democratic deficit to make use of the political opportunity of the 2002 general election. This process of amplification, transformation and generation of meanings around GSE's three main frames occurred as part of what Snow and Benford (1992, p. 141) have described as the 'cycles of protest.' This study has highlighted the link between GSE's collective action frames and the cycles of their campaign against the states policy of pushing incineration as a waste management option.

8.2. MOBILISATION OF INTERNAL RESOURCES

This study has examined the GSE case by placing it in a number of contexts. In the theoretical framework presented in Chapter 2, interpretations of environmental social movements are presented, to contextualise the type of movement analyses used to rationalise GSE's case. Similarities with social movement activities are evident in GSE's mobilisation of internal and external resources (McCarthy & Zald, 1977; (Tarrow, 1994, 1998). GSE organised itself in a manner which is congruent with understandings of new social movement behaviour (Martell, 1994). Its campaign was based on an 'anthropocentric' interpretation of an environmental rights issue (Eckersley,

1992), as concerns developed about the potential health risks posed by the incinerator proposed for the Galway area as part of the *Connacht Waste Plan* in late 1999 and early 2000. This anthropocentric or human-centred perspective on environmental issues is at the heart of NIMBYist campaigns, and this view motivated GSE at its inception.

GSE relied on the expertise of 'new middle class professionals' to mobilise the internal resources available to it (Inglehart, 1977; Melucci, 1996). GSE conformed with established understandings of environmental movement activity when it exploited the 'shared grievances' (Klandermans, 1989) of a concerned public, based on health risks. In so doing, GSE displayed movement characteristics similar to those highlighted by Scott (1990) who argues that movements may wish to increase access by exploiting issues of environmental concern. Both Scott (1990) and Offe (1987) have argued that mainstream political parties have difficulty challenging the 'technocratic criteria' for policy that is a feature of neo-corporatist states such as the Republic of Ireland (Roche & Cradden, 2003). Scott (1990) argues that this arrangement can stimulate environmental movement challenges. The GSE case is an example of this form of new middle class environmental response to such neo-corporatist exclusion, as it challenged the technocratic nature of the state's waste policy as part of their democratic deficit frame.

8.3. IRISH ENVIRONMENTAL MOVEMENTS

Comparative cases of other anti-incinerator campaigns in Ireland and the United States have been included to provide an insight into previous attempts by environmental movements to gain access to political opportunity structures (see also Leonard, 2006, 2008). According to writers such as Tovey (1999), Allen (1990) and Baker (1990) Irish environmental movements could gain some political access without being able to influence policy making. The pretext for the degree of closure, faced by environmental and anti-toxics movements was based on the state's prioritisation of job-creation and economic growth during the decades of the 1970s and 1980s. The directives of the EU also increased pressure on government in regard to Ireland's increasing waste management problem. Forced to introduce a waste management regime due to European Commission directives, the state established what has been described as a pro-business Environmental Protection Agency (Taylor, 2001). This study has presented two stages of environmental campaigning, a pre-boom anti-toxic multinational phase followed by a post-boom anti-waste infrastructure phase.

The second phase campaigns have been characterised by NIMBYism and localism.

One theorist on Irish environmental movements has claimed that Irish environmental movements have suffered due to 'the failure to connect' (Cox, 1997, p. 40) to other like-minded movements. My research indicates that GSE's ability to utilise the technologies of communication and forge links with wider movements was a significant aspect of their case, one that had not been fully exploited by previous movements. GSE's primary success in this regard reflected its ability to use Internet resources to build a relationship with its primary scientific source, Paul Connett.

While Cox has argued that community movements have 'local rationalities' based on the concerns of 'ordinary participants' (Cox, 2001), this study has indicated that GSE's campaign was led by professionals with the expertise and networks necessary to move beyond its original local concerns. The mobilisation of the resources of expertise and networks provided GSE with political opportunities in the internet era, which had been unavailable to Irish environmental movements in previous disputes. The access to this information was provided through its links with Connett. This utilisation of communication technologies allowed GSE to create networks that produced moments of 'increased access' (Tarrow, 1994, 1998) that were not achieved by previous campaigns. This ability to use media to gain momentary access at key stages of the campaign provided GSE with a degree of success at key moments. These key moments of access included their influencing of the council's rejection of the CWP, and their success in having all parties (except Fianna Fail) removing any support for incineration from their 2002 election manifestos.

The recurring role of experts, such as Dr. Paul Connett, in Irish environmental disputes, as well as links that were forged between the anti-incinerator groups, as well as the Cork Environmental Alliance and the DuPont protestors in the 1980s, indicate a degree of networking that goes beyond a concern for 'local rationalities' (Cox, 1999b). While Cox (1997) decries the 'weakness of movement culture', due to 'the failure to connect' (*ibid.*), with other movements, my study argues that the networks established at various moments in Irish environmentalism allowed a movement culture to be sustained. It was this culture that provided GSE with the context for its campaign. Through their 'interpersonal contacts' and utilisation of communications technologies, GSE was able to frame a 'moral discourse' (Grove-White) around its perspective on the incineration debate, linking with US experts and Irish movements in a manner that achieved 'increased access' to POS (Tarrow, 1994, 1998) at certain points of their campaign.

However, while the ability of Irish movements to create limited networks has been established by this study, the shift from what Cox (1999b) has described as 'local rationalities' cannot be said to amount to the movement transition described by Szasz as 'an ideological development from NIMBYism toward a radical environmental populism' (Szasz, 1994, p. 69). While Irish environmental movements have been described as 'populist' (Tovey, 1992), this study contends that they are often reliant on the new middle class who have the professional expertise necessary to create movement consensus and mobilisation. In this way, Irish environmental movements can be said to be driven by the 'issue-driven' (Grove-White, 1993) professionals who maintain the 'interpersonal contacts' necessary to provide an ongoing pool of resources that subsequent movements can draw on over time.

The significance of this resource network for movements that are attempting to gain access at various stages of the evolving political opportunity structure cannot be underestimated even though Fagan et al. (2001) make a point about the depth of local feeling created by waste disputes, their division of local responses into 'negative' (NIMBYist) and 'positive' (environmental) fails to realise that a path from one to the other may be put in place, as the GSE case indicates. This study demonstrates that the GSE case can be placed somewhere in between, as an issue-driven campaign that mobilised beyond its NIMBYist inception, but that failed to make a complete breakthrough towards becoming integrated in either the radical or formal political sectors. In this regard, GSE was a NIMBYist group that embraced ecopopulist agendas through professionally led public and media campaigns.

Another study that underplays the onset of the networking achieved by Irish environmental movements is that of Garavan (2004). He relies on one primary source, the *Irish Times*, as well as a meeting of twelve activists at the Convergence Workshop in Dublin in 2002, to conclude that environmental activism in Ireland has a 'local' focus (Garavan, 2004, p. 69), with 'low levels of resources' (*ibid.*, p. 14). These findings can be disputed, in the light of the evidence of the mobilisation of the resources of networks and expertise revealed in the GSE case, particularly in relation to the mobilisation of 'interest-led' science through their use of internet technologies, and their links with Connett, CHASE and the NIA.

While my research has established the significant influence of foreign experts such as Dr. Paul Connett, to several environmental campaigns in Ireland, such international networking is not referred to by Garavan. Instead, he reports that 'the international networking of the Irish environmental movement appeared to be ... very low' (Garavan, 2004, p. 103). While the extent of the wider networks established by GSE could be said to be moderate, the link with Connett was significant to their campaign, and an attempt was made to forge links with anti-incinerator campaigns in Cork and Meath.

Garavan's study doesn't acknowledge the significance of the utilisation of communication technology for Irish environmental movement's networking with US experts such as Connett, something that has been established as a critical feature of GSE's campaign by my research. Furthermore, the moments of 'increased access' (Tarrow, 1994, 1998) achieved by environmental campaigners such as GSE is not recognised by Garavan, who instead argues that 'the government is under little popular pressure to accede to environmental demands' (Garavan, 2004, p. 109). Garavan argues that the Irish environmental movement hasn't impacted upon the politics of the state. Conversely, my research indicates that the populist pressure of NIMBYist groups, such as GSE, CHASE and NIA did amount to a challenge to state policy, delaying waste management projects. In the second phase of campaigns, the state has come under increased pressure due to EU directives and increased waste. Garavan also argues that movements are 'severely limited in their capacity to act through a lack of material sources' (*ibid.*, p. 210). However, my research has established that movements such as GSE were innovative in their ability to mobilise resources such as expertise and networks, and that these resources provided 'increased access' (Tarrow, 1994, 1998) at certain points to the political opportunity structure. Due in part to the timing of their campaign, GSE were able to mobilise professional expertise and communication technologies, as well as establish an anti-incineration internet site, in a manner that was unique to Irish environmental movements up to that point.

While literature on Irish environmentalism has focused on the importance of legal challenges and oral hearings as a strategy for community campaigners (Taylor, 2001), this research provides evidence of an environmental movement that achieved some access to the POS by avoiding the expensive and time- consuming legal route, moving from public and media campaigns to formal politics, during the 2002 election campaign. GSE attempted to exploit the 'unstable alignments' between the coalition parties who were competing for seats in Galway West (Tarrow, 1994, 1998). While this strategy didn't provide GSE with the 'influential allies' (*ibid.*) it hoped to have at the cabinet table, it has at least led to the implementation of the state's waste policy being delayed. While announcements about municipal incinerators have recently been made in Meath and Cork, no such announcement has been made in Galway as of June 2005. In this context, GSE's reliance on a tactic of influencing electoral politics, rather than going down the more established legal route indicates a departure from established approaches to environmental campaigns in Ireland.

The political leverage that GSE attempted to achieve through its campaign to influence the 2002 general election could have paid off. This potential for the use of environmental issues in electoral politics is not dealt with by the existing literature on Irish environmentalism (Allen & Jones, 1990; Taylor, 2001; Tovey, 1992). While the Irish Greens have had electoral successes (Mullally, 1997) they haven't always maintained links to grass-roots movements at the key moments of 'increased access'

While Taylor (2001) identifies the dilemma faced by 'green fundamentalism' (*ibid.*) when it 'questions the role of science' (*ibid.*), GSE's mobilisation of the resources of expertise and communications technologies to provide a viable form of 'interest-driven science' with which to challenge 'official science' (Grove-White, 1993) did provide a degree of leverage that 'increased access' (Tarrow, 1994, 1998) at key points of their campaign. The opposing scientific views in GSE's case indicate that all science may be 'interest-driven', with favourable data being adapted by campaigners and the state to support their own positions. Ultimately, GSE's ability to mobilise internal resources and exploit external political opportunities provided the momentum for their challenge to the state's waste policy, with its temporary and contingent increased access leading to mixed results for their movement.

The state's waste policy has resulted in community protests while the state's attempts to address the bourgeoning waste crisis that has emerged in the aftermath of increased consumption associated with accelerated growth. Ireland had fallen foul of European directives on waste management, due to an over reliance (of up to 93%) on landfill. To deal with this crisis, waste policies were introduced with plans for regional incinerators across the country. Previous attempts to introduce an all-island incinerator in Northern Ireland had been opposed by both nationalists and unionists together. As a result of the networks created from the Northern campaign, communities mobilised in small regional cities and more latterly in Dublin (*ibid.*).

These regional campaigns benefited from links with the global anti-incinerator and 'zero-waste' campaigns, and from the support of experts from the United States and Britain. As a result, the state's regional waste policy was challenged, as communities began to inform themselves of the dangers posed by the dioxins and furans which are released into the air and food chain from incinerator emissions. Nonetheless, there were positive outcomes to the community engagement with the waste issue, and regional recycling was introduced in many areas with great success. The state has not abandoned its plans to introduce incinerators completely however. Two major plants are planned, and these will deal with the bulk of municipal and industrial waste.

Ultimately, the costs of economic growth have led to greater environmental degradation and an undermining of significant aspects of the nation's heritage. In particular, rural communities have borne the cost of Ireland's transition from a rural agrarian society into an industrialised and globalised economy. While communities have demonstrated a commitment to retaining some aspect of the traditional values and heritage of the 'Emerald Isle', the state's policy of prioritising rapid infrastructural development in order to underpin private consumption and multinational investment over local community services may prove to be ill advised as fluctuations on the global markets continue. As the cycle of the 'Celtic Tiger' draws to a close, the values and concerns of these local communities may provide a new direction for the post boom era.

There are of course many competing understandings surrounding community-based environmental disputes in Ireland with almost as many accounts as there are participants or advocates who often wouldn't always see themselves as 'environmentalists'. Environmental scientists also take a different viewpoint calling for accuracy and expertise during debates about the anthropocentric development of the landscape while the technocratic advisers employed by the multinationals or the state always find enough evidence to demonstrate the safety of their technology no matter how many complaints such technologies have raised in the past. Therefore, we can understand many community-environmental disputes in Ireland primarily as debates about rational choice. For communities the scientific evidence which demonstrates the safety of technologies provides little comfort when prior campaigns have argued about the inherent risks posed by the same technologies. And so many disputes become contests between rational science versus community concerns with one side marshalling data and the other mobilising grievance. Set in this context it seems difficult to imagine that communities bring themselves to accept infrastructure or technologies until after they have been proven to be safe. Such is the nature of the 'risk society' outlined by Ulrick Beck (1992).

POSTSCRIPT 1: THE GALWAY WATER CRISIS

Galway provided another example of 'risk society' with the outbreak of a parasitic-related contamination of municipal water supplied in 2007. The 'Galway Water Crisis' emerged in March of 2007, in the aftermath of an outbreak of the cryptosporidium parasite in the local water system.[1] This crisis reflects the failure to protect large bodies of water such as Lough Corrib from the impacts of human development. As the degradation of water supplies has continued, urban centres such as Galway have had to contend with boil notices, health warnings and a political 'blame game' in the run-up to the 2007 election. This chapter will examine the key issues surrounding the water crisis in the west, detailing the costs of this issue to those charged with dealing with it.

One offshoot of the accelerated growth which occurred in the Republic of Ireland since the mid-1990s has been an increase in crisis related to infrastructural development. While levels of personal wealth have increased dramatically, the accompanying growth in consumption and waste has created a series of crisis and subsequent protests about sewage treatment plants, landfills or planned incinerators, gas pipelines, roads, electricity pylons and mobile phone masts. In many cases, a political campaign of opposition has emerged from communities concerned about the risks associated with such developments (Leonard, 2005, 2006). However, in the case of the cryptosporidium outbreak in Galway in the spring of 2007, a 'blame game' of accusation and counter accusation broke out between the various layers of institutional politics involved in the provision of municipal water to the city and its environs. This political framework included a diverse range of actors including the unelected technocrats of both the city and county councils in Galway, the elected councillors of both institutions including the city's Mayor Niall Ó Brolcháin of the Green Party and the Minister for Environment in the Coalition Government, Fianna Fáil's Dick Roche. This chapter will examine the extent to which this issue became part of the wider political agenda of electoral politics in the run-up to the 2007 general election, both within the local context of the campaigns of the parties in the Galway west Constituency but also as an underlying factor in the formation of potential coalitions for government subsequent to that election's outcome.

Of course, some scientific definition of this intestinal infestation is required at this point. The cryptosporidium bug is a microscopic parasite which becomes attached to the human intestine when ingested. It results in cramps and diarrhoea for most people, with severe cases of infection in very young or old people leading to severe illness and in a few reported cases, even death. The key source of the bug is both animal and human waste, with any infestation of water supplies usually being attributed to waste appearing in the water supply at some point (*Irish Times*, 14 April 2007). The issue of water pollution goes back some time before the current crisis. Anglers on Lough Corrib have had long-standing concerns about water quality, while scientists attempted to raise the issue as far back as 1997 when the Geological Survey of Ireland first highlighted a cryptosporidium risk for the Galway region due to the absence of soil layers prevalent in other parts of the country. In the west, the indigenous karst limestone topography provided little in the way of the filtering effect necessary to break down micro organisms such as cryptosporidium, increasing the risk of water contamination (*ibid.*). In the case of Lough Corrib, the largest lake in the Republic, effluence from large tracts of counties Mayo and Galway is washed through the lake and into the water supply for most of the local population, including the 70,000 inhabitants of Galway City. Towns surrounding Lough Corrib and its tributaries such as Oughterard and Headford have become part of the bourgeoning commuter belt around the prosperous city of Galway, enticed by the promise of profit from selling up in the city which is second only to Dublin in second hand property prices. As a result cheaper housing developments have sprung up around all of the towns near Galway city, replicating the inflated national property market which has been synonymous with the second phase of the 'Celtic Tiger' once its original determinant of multinational investment had subsided in the years after the 9/11 attacks. The fact that Galway had maintained its multinational presence through large multinational employers such as Boston Scientific and Medtronic meant that a steady flow of people continued to be attracted to this western where its two third level institutions and cultural reputation based on the successes of theatre groups such as Druid and Macnas provided Galway with the reputation of a dynamic young and growing city.

It was during one such cultural event, the Macnas inspired St. Patrick's Day Parade in March 2007, that news began to spread of a contamination threat to the city's water supply. Tests by the Health Service Executive (HSE) were conducted across the city and county as the extent of the problem began to emerge. Residents in the city and surrounding areas were promptly advice to boil all drinking water until notified otherwise. Panic buying of bottled water was reported over the bank holiday weekend as hotels, bars and restaurants struggled with the influx of tourists for the St. Patrick's festival. The boil notice was announced as a precaution while the authorities searched for the source of the outbreak. Testing was initially concentrated at the city's ageing Terryland waterworks which were first constructed in the late 1940s with a second facility being built in 1972. Most municipal waterworks contain the cryptosporidium bug through a filter system. However, the older Galway facility had a simple mesh to remove large pieces of waste, followed by a process of chlorination which in itself doesn't kill the cryptosporidium bug. The local council was aware of a significant risk of cryptosporidium from 2005 (*Irish Times*, 14 April 2007).

This news was enough to engender an immediate response from the government, particularly in an election year. Fianna Fáil's Environment Minister Dick Roche had spent months promoting himself as a 'green' politician, particularly in the wake of opinion polls which indicated that a Fianna Fáil/Green coalition was feasible after the next general election. However, this green façade was soon removed when Minister Roche opened up his own front response to the water crisis, accusing Mayor Ó Brolcháin of incompetence on the issue:

> We're about to go into the season when Galway becomes a very important tourist Mecca so there is a huge amount resting on this issue. It will take some time to get a new waste treatment plant built and we have to resolve the matter in the interim. A mistake has been made and there certainly wasn't the degree of urgency that should have been put into this issue. What we must do now is ensure that the urgency is injected into it. (*Environment Minister Dick Roche RTE? News*, 29 April 2007)

One issue that was clarified by the tests conducted by public health scientist Dr. Diarmuid O'Donovan was that the source of the crisis was human rather than animal waste. This confirmed the fact that unregulated development had been the cause of the water contamination due to the insufficient waste treatment plants in areas that had been overdeveloped along Lough Corrib. However, for the beleaguered citizens of Galway, the cause of the crisis was of much less significance than the solution. Within weeks of the crisis, a populist community movement was mobilised in response to the issue. This movement was called simply 'The Galway

Water Crisis' campaign. On their web page, this group explained who they were:

> A number of people are asking 'who are you?' So, just to clarify, we are ordinary citizens of Galway. We're not a branch of any group, cause, organization or political power. We welcome all people regardless of age, ability, gender, creed, colour, ethnicity, minority group, work status, sexual orientation, political persuasion or organizational affiliation. All are welcome.
>
> We all need water. (www.myspace.com/galwaywatercrisis)

The Galway Water Crisis campaign stated that the campaigners as a group felt 'let down' by both local and national politicians who 'seemed more interested in passing the blame than actually doing anything about this crisis' (*ibid.*). Kelly's letter continued by stating that the state was taking one and a half million euros a week in VAT for water sales, and that the group wanted that money used in the efforts to solve the crisis. Labour councillor Billy Cameron criticised the proposed use of UV technology, stating that ultra violet tubes were less efficient in water which had darkened with a tint, as the contaminated water in the Corrib had due to pollution. In addition, he claimed that while UV light worked with most microbes, it wasn't effective on protozoa such as the cryptosporidium. The councillor suggested a Russian method of electrochemically activated water or anolyte, which was a non-toxic way of destroying all microbes effectively. Councillor Cameron claimed the crisis arose as a result of 'a government which is absolutely development led' (*Connacht Sentinel*, 8 May 2007). One Fine Gael candidate for the Galway East constituency, Dr. John Barton, felt that culpability rested with a number of groups such as the HSE, the city and county managers, the elected councillors, the Department of the Environment and the Environmental Protection Agency. Dr. Barton stated that risk assessment procedures were not carried out properly over many years, and that the exceptionally heavy rainfall and subsequent flooding which had occurred in Galway that winter should have alerted the authorities to the need to monitor the water supply for problems (*ibid.*). One editorial claimed that gallons of raw sewage were being pumped into the rivers and lakes of Galway, as well as Galway Bay due to new housing developments. The editorial called for an audit of all water supplies in the county, castigated Minister Dick Roche for trying to shift the blame for the crisis when 'he should bear the final political responsibility' (*ibid.*).

For their part, Galway city manager Joe McGrath claimed that the water crisis would end on June 15 with the increase in supply of clean water from the Luimnagh plant. The council also had plans to repair water pipes in the city. However, with the general election taking place at the end of May, the local media began to speculate as to which candidate would be most affected by the water crisis. The issue had relegated all others to secondary importance, becoming the main election issue in Galway West constituency. One candidate, councillor Brian Walsh of Fine Gael, stated that the issue was being raised on 'at least 50% of all doors we canvass' (*Galway Advertiser*, 26 April 2007). The Galway Water Crisis campaign had found large levels of public anger with public officials due to the crisis. Speculation began to emerge as to which of the city candidates the crisis would most affect amongst a diverse group including Mayor Niall Ó Brolcháin of the Greens, Michael D. Higgins of Labour, and councillors Brian Walsh (FG), Michael Crowe and Frank Fahey of Fianna Fáil, Noel Dempsey and Donal Lyons of the Progressive Democrats and Independents Catherine Connolly and Margaret Cox, who had broken away from Labour and Fianna Fáil, respectively, after not being selected to run for those parties. Some commentators felt that Ó Brolcháin, as Mayor, hadn't shown strong leadership on the issue, while others felt that the Fianna Fáil candidates such as Crowe and Fahey would lose out due to the crisis. The independents and opposition parties also had to contend with the campaign of the Galway Water Crisis group, who claimed to be non-party political. The group claimed that public faith in the political system had been 'irreparably damaged' by the water crisis. Many felt that the risk of an outbreak of cryptosporidium even after a new plant was built would remain due to poor monitoring and unregulated development, according to the group:

> The feedback we are getting from the public is that they would be afraid to drink the water from the city system again. Unless we tackle the major problem of cleaning Lough Corrib, then this problem will simply come back again and again. (*ibid.*)

Maeve Kelly of the group claimed that the problem was one which affected all the towns around Lough Corrib, not just the city areas reliant on the old Terryland plant. These areas had unreliable sewage systems which pumped effluent into the Lough. For Kelly, 'the major problem lies with the cleaning of the Corrib. There will always be a problem as long as raw sewage is coming into the Lough' (*ibid.*). The water campaign brought its protest to a meeting of Galway City council, calling for 'Clean Water for Galway'. Protesters brought their empty water bottles to deposit at City Hall as part of their protest, to highlight the rising cost for citizens having to buy water and recycle bottles, in a city which ironically had led the way when recycling pilot schemes had been first introduced (Leonard, 2005). The group set up a web page at www.myspace.com/galwaywatercrisis, which had nearly 1,000 hits in its first two weeks. For those candidates who had cut their political teeth as environmental advocates such as Niall Ó Brolcháin (anti-landfill) or Catherine Connolly (anti-incinerator), their new roles as city councillors or mayors created constraints as they couldn't campaign against the council which they now represented. This issue of co-optation of former advocates would become apparent during the 2007 general election campaign. For its part, the Galway Water Crisis group continued with its campaign of public meetings, protests and petition gathering, tactics which had served Ó Brolcháin and Connolly well during their time as social movement campaigners against incineration during GSE's campaign (Leonard, 2005, 2006). Senator Margaret Cox, who would break away from Fianna Fáil to contest the election as an independent, called for better information for the public on the issue. The experienced President of the Labour party and election candidate Michael D. Higgins called for localised filtering systems so people could collect free water. He claimed that the Government had plans to privatise water supplies across Ireland, something which led to the delay in 'the allocation of EU and state funds for water and sewage services to local authorities' (*Irish Times*, 24 April 2007). Higgins pointed out that while citizens faced payment of large fines from the European Union due to the non-implementation of European Union Directives, they were also being denied clean water due to 'government failure' (*Galway Independent*, 25 April 2007).

The Mayor did receive some support from An Taisce, who claimed that the construction of new housing should be frozen until the water crisis was sorted out. This moratorium should remain in place until 'the essentials of life' such as good drinking water, sewage treatment and waste disposal systems were sorted out (*ibid.*). The fact that national polls were indicating that the Greens could become numerically viable coalition partners for Fianna Fáil after the votes were tallied seemed to increase the pressure on Ó Brolcháin. However, when Ó Brolcháin did choose to fight back, he targeted Fianna Fáil's sitting TD Frank Fahey and the issue of the proposed city bypass, rather than raising the water crisis. This tactic backfired on the Mayor, as the city's traffic faced deadlock and the promise of a light rail system was also prominent in Fahey's own campaign literature. Furthermore, the Green Party leader Trevor Sargent had also targeted both Fahey and party leader and Taoiseach Bertie Ahern through investigations into their financial matters, which ultimately proved fruitless. In addition, the Greens left the PD's Noel Grealish with a clear run in the Carnmore/Oranmore area which had seen a considerable population growth in recent years. Moreover, Fianna Fáil were always hopeful of regaining three seats in

Galway West, and were therefore unlikely to allow a second seat to be lost in the constituency, reducing them to one sole representative, Eamon Ó Cuiv in Connemara. In these circumstances, it may have been more expedient for the Greens to target Grealish's seat more keenly than Fahey's, particularly as the Progressive Democrats were doing extremely poorly in national polls. In fact, Fianna Fáil appeared to anticipate the swing to the Greens over environmental issues such as climate change, with Michael Crowe's election literature being devoted to green issues. At this point, party leader Bertie Ahern attempted to stake a claim for the young green voter when speaking at the Fianna Fáil youth conference in Galway during the election, a week after the Green Party's own national conference was held in the city. Commentators claimed that Ó Brolcháin would have to receive enough first preferences to retain a chance of being elected, as the transfer of votes would be too unpredictable in Galway West. However, when the Greens held their main press conference of the election, Ó Brolcháin's statement that the water crisis was due to bad planning and a lack of resources for the extra staff seemed to some commentators like a bureaucratic excuse, rather than presenting any solutions to the issue. For instance, the issue climate change, which caused the initial problem due to the floods, wasn't mentioned. In addition, his unqualified opposition to the Galway City bypass seemed contradictory in comparison to his support for bypasses in nearby towns such as Claregalway, Loughrea and Ennis. Reports that Galway City Council was considering water rationing and pressure reduction didn't help with the public's mood in the week of the election.[2]

Invariably, the water crisis was one of the factors put forward by Ó Brolcháin as causes for his failure to win a seat in a constituency where many had thought he would be successful. The crisis was one of the key issues which damaged Ó Brolcháin's chances, alongside the bypass and the national swing to the mainstream parties of Fianna Fáil and Fine Gael. However, the success of the Progressive Democrat's Noel Grealish left no room for Ó Brolcháin, as Galway West returned its five sitting representatives. Ó Brolcháin was one place ahead of fellow activist turned councillor and mayor Catherine Connolly, who polled well in her local ward but failed to make any inroads into the vote of long-standing Labour candidate Michael D. Higgins. The political opportunity structure surrounding both of these activists turned politicians had become constrained, as the formality and expectation of mainstream politics prevented them from campaigning on issues such as the water crisis with any effect. Charged with solving the problems of the bourgeoning 'City of the Tribes', neither Ó Brolcháin nor Connolly could distance themselves from the public's own desire to play the 'blame game' in the aftermath of the water crisis. In many ways, this predicament displays the dilemma facing activists who become co-opted by the mainstream, where the potency of collective action is sometimes traded for the robes of office. Ó Brolcháin's loss was heightened when the Greens struck a deal with Fianna Fáil to enter a coalition government after the election. He claimed that holding the office of mayor and the negative campaigning on the water crisis had been part of the reasons for his electoral defeat:

> Being the mayor while there is an election campaign on is probably more of a disadvantage ... it meant I had to concentrate on being mayor and was not able to devote the sufficient time to campaigning ... it (the water crisis) came at a bad time for me as mayor, because as one person put it on the doorstep, 'well logically we don't blame you, but there is no one else to blame so we have to blame you'. Because there is a degree of responsibility being the mayor ... I felt it was important to try and sort out a lot of problems rather than being overtly political about it. (*Connacht Sentinel*, 29 May 2007)

While Ó Brolcháin waited for an offer to enter the Seanad, the original deadline of the 15 June for the reopening of clean water supplies for Galway came and went. As the summer tourist and festival season appeared under threat from the ongoing crisis, Galway's newly returned TDs began a post election assault on the city management for failing to meet the target date for clean water in the region. With the election now over, no one was blaming the mayor anymore. The Galway Safe Water campaigners called for the sacking of the city manager, a call which was welcomed by Ó Brolcháin: 'I would welcome a prosecution; I think the finger needs to be pointed' (*Galway City Tribune*, 15 June 2007). On Monday 18th June 2007, Labour councillor Tom Costello was elected the new mayor of Galway. Niall Ó Brolcháin's journey from environmental activist to Green representative had come full circle. The problem of co-optation for activists who embrace the political mainstream remains an outstanding obstacle for those environmental advocates now faced with negotiating with a government which includes a Green party Minister for the Environment. The beleaguered citizens of Galway had little respite through the summer festival season. The HSE announced that up to 238 cases of cryptosporidiosis had been recorded in Galway in 2007, with non-reported cases estimated in much higher numbers. The boil notice was lifted in east Galway in July, due to the extension of services from the Luimnagh plant (*Galway Advertiser*, 21 July 2007).

The famous Galway Races Festival was declared cryptosporidium free due to a 100,000-euro investment in water filters at the Ballybrit race course. The Macnas street theatre group announced it would use only bottled water as part of its water-themed parade, which featured a giant cryptosporidium 'bug' (touted as 'the ninth wonder of the world') as one of its main characters. Commercial interests, which had complained of losses during the tourist season, were offered a 10% rate rebate. In a summer when nearly a quarter of a million lost water supplies in the south western region of England due to extreme flooding, Irish towns such as Ennis and Clonmel were also hit with boil notices due to flooding. For Galway's local authorities, the loss of confidence from the public in its services was equalled by costs of over 400,000 euros in weekly laboratory tests at its water plants. After a series of delays, water supplies to the city and its western environs were returned to Galway city as the last obstacles to providing acceptable levels of clean water from Luimnagh to the city were overcome (with supply rising from 14,000 m^3 of water to 17,000 m^3), according to City Manager Joe McGrath (*Galway City Tribune*, 13 July 2007).

With cryptosporidiosis cases dwindling to pre-crisis levels, the Galway water debacle was coming to an end. Water supplies were restored on 20th August. The final bill for the crisis was set at 15 million euros, including the costs to industry and private households (*Connacht Sentinel*, 21 August 2007). However, with reports of increased instances of extreme weather forecast by the United Nations World Meteorological Organisation (WMO) after a year when extreme weather events occurred in a 'range which is well outside the historical norm' (*Irish Times*, 9 August 2007), the threat of instances of flood-related water contamination will remain for communities in Ireland and across the planet. With poor infrastructure and a culture of political blame, citizens across Ireland will be hoping for a sustained and effective response from their political leaders to address this issue before another crisis occurs, particularly in the heavily populated east coast. However, for the Irish Greens, the 'blame game' proved to be descisive, and the party lost virtually all their seats in the 2009 local and 2010 general elections.

NOTES

1. See also Leonard, L. (2010). The Galway water crisis of 2007. In M. Corcoran & P. Share (Eds.), *Sociological Chronicles 5*. Dublin: Liffey Press.

2. The 25th May 2007 election results for Galway West were as follows: electorate: 86,602; total poll: 55,627; turnout: 64.24%; quota: 9,183; Ó Cuiv, Eamon (FF) 9,645 (elected); Higgins, Michael (Lab) 6,086 (elected); Fahey, Frank (FF) 5,854 (elected); Grealish, Noel (PD) 5,806 (elected); McCormack, Padraic (FG) 5,419 (elected). Not elected: Crowe, Michael, (FF) 4,969, Healy Eames, Fidelma (FG) 3,904, Ó Brolcháin, Niall (GP) 3,026, Connolly, Catherine (IND) 2,006 (www.rte.ie/election 2007).

POSTSCRIPT 2: ECOLOGICAL MODERNISATION AND IRISH ENVIRONMENTAL POLICY

ECOLOGICAL MODERNISATION

Introduction

The increasing trend towards sustainable development has seen a shift from 'end of pipe solutions' to the ongoing threat of pollution. Policy makers have come to accept the need for some form of inbuilt environmental standards to be included in any overall planning strategy. These shifts come in the wake of the *Brundtland Report* and the Rio World Summit. They have also shaped environmental policy. A central feature of this new thinking is the theory of 'Ecological Modernisation' (EM). Underpinning this debate are the theorists Janicke, Weale and Hajer, who have each contributed to the conceptualisation of EM as a feature of modern society. It can be argued that EM theory reflects a critical new positioning of the environmental debate, moving away from the periphery of social, cultural and political channels and becoming an important aspect of policy planning in these areas.

There has been a marked increase in trends towards the restructuring of Western economies with a view to improving environmental standards. Central to this strategy is the concept of EM, first identified by the German writers Janicke and Hajer in the early 1980s. EM is based on the idea of sustainable development becoming a central feature of future policy processes; to do this ecological factors must become an integral part of social, cultural and industrial planning. Environmental efficiency and lifestyle benefits should be highlighted as the ultimate rewards for embarking on new, eco-modernist approaches to national and global problems.

However, EM has become the subject of criticism from many of the more radical environmental groups who fear that EM will be used as a 'catch-all' for ecological policy while, in reality, many issues on the environmental agenda remain unresolved. One result of this debate has been the spread of shifting ideologies within the movement out of which the 'light' and 'dark' green camps have been established. Within these new paradigms an accommodation of environmental concerns must be identified from the wide range of actors involved including those in wider society sympathetic to environmentalists' ideas. The state agencies and corporate institutions who shape and act on environmental policy must also use these new paradigms.

With this in mind ecological activists argue that policy must be measured in relation to its impact on both local and global situations. If EM holds out the promise of a step forward from an existing ecological consciousness and as a detriment factor within active society, rather than as a peripheral player, it can only do so by confronting the dichotomy between these two apparently irreconcilable opposites.

Beck has written about and led an examination of EM's role in the ecological movement. Janicke has proposed a theory that a stable democracy is a vital condition of an environmental policy being implemented. Hajer presents a discourse in relation to institutional change and Weale investigates how a new political approach has grown around the ecological agenda.

EM, Sustainable Development and the Bruntland Report

Ecological modernisation is an alternative conceptual approach to what were previously seen as costs and benefits of effects on the environment and is presented as an integral part of the planning process. Costs were shown to be beneficial as they would outweigh future costs of repairing the damage caused by pollution. This approach coincided with a new, consumer-driven demand for eco-friendly goods that arose in the 1980s. At a policy level EM has been used to help forge new and coordinated approaches to what were seen as problems facing the differing strata of bureaucracy. Inefficient approaches to these problems led to a crisis in legitimacy and regulatory failure. The eco-modernist discourse has provided a fresh approach or what Weale calls a 'new politics of pollution' aimed at harmonising previously competing actors with an improved societal arrangement as the reward.

Recent debates on ecological approaches have been influenced by an acceptance in certain political and economic quarters that free-market forces, left unchecked, could damage the environment in a costly and permanent way. Following analysis in the 1970s, such as the paper issued by the Club of Rome on the 'Limits to Growth' a report titled 'Our Common Future' was released by the World Commission on Environment and Development (WCED). The paper became known as the 'Brundtland Report'. In its definition of sustainable development the Brundtland Report describes the need to implement plans for 'development that meets the needs of the present without compromising the ability of future generations to meet their own needs' (Our Common Future, 1987, p. 43). Its objectives were to become a crucial factor in ecological debates. It included policy measures such as reviving growth, meeting job levels linked with food, energy, water and sanitation needs, maintaining population figures at sustainable levels and most importantly from an environmental perspective, linking economic and technological planning as an integral part of any policy development. This relationship has become an important aspect of ecological theory and has had an impact on environmental thinking.

Questions of Legitimacy in Democratic Institutions

In order for environmental policies to be successfully implemented a stable democracy must be in existence. To support this argument one could cite the failures of right wing dictatorships such as Portugal, Spain, Greece and former Eastern Bloc Countries on maintaining any form of pollution control. Central to this fact is the dichotomy between economic expansion and the need to restructure social mores around ecologically friendly strategies. The list of criteria vital to the Green movement, if it wishes to make serious gains on the policy front, was set by Janicke (Janicke, 1999). His list includes a pluralist, democratically elected parliamentary system, civil rights, free media and an enquiring, well-funded science programme. By contrasting the environmental performance of Western democracies with the former authoritarian states Janicke provides the evidence which suggests strong links between factors of wealth, gross national product (GNP) per capita and success in policies dealing with such problems as sulphur dioxide emissions or attempts at recycling. Studies also point to the fact that while developed nations' higher standard of living created more pollution, these states invariably have the means of dealing more efficiently with such problems. Links are also found in standards of education and resulting lifestyle changes which create an environmentally minded consciousness within these sections of society.

There is little evidence of policy achievements in environmental matters being related to left wing, liberal or conservative parties being in government. However, these governments would acknowledge the importance of the European Parliament's (EU) overall environmental agenda. This emphasises the need for an open policy process so that small parties and non-government operatives (NGOs) have access. Strong information systems and ongoing education are crucially important to policy successes. Studies have also shown the need for well-structured ecological pressure groups and full-time environmental agencies, referred to by Janicke as 'a kind of eco-corporatism' (Janicke, 1999, p. 75). Janicke also emphasises the findings of repeated studies which underline the need for cross-sector policy cooperation and to establish links between the various strata of corporatist bureaucracy. In effect, these studies detail the knock-on benefits of creating a culture of consensus politics in which the ensuing atmosphere of shared aims and goals becomes conducive to the creation of an ecologically minded consciousness at an institutional level.

It is the parameters of open political structures coupled with freedom of information which underpin much of what Janicke considers the necessary criteria for successful EM policy with EM positioned within the actual dynamic of social structures rather than as an agitator on the periphery. Technological and bureaucratic efficiency becomes central to the creation of any new society that wished to overhaul its environmental performance. Society is called on to reclaim some of the responsibilities which have been left to a cumbersome central government with the elimination of wasteful production being sold to industry as an indicator of ecological and efficient productivity while new pressure groups and a global media take on roles as environmental watchdogs beyond the nation state's capacity. This emphasis on a civic participatory response to the crisis of legitimacy experienced by Western governments is central to Janicke's views on World Environmental Capacity (Janicke, 1999) in the face of regulatory failure.

EM theory has been built around the view that political modernisation comes through society's response to and capacity to deal with the crises it encounters. From an environmental point of view the response by Western nations to the oil crisis of the 1970s and the corresponding economic slump led to the questioning of growth and consumption levels and ultimately opened the door at policy level to environmentalists who had previously been seen as an extreme group on the margins. Environmentalists placed a great emphasis on the need to increase levels of civic participation while simultaneously presenting an integrated departmental response from government. Access to planning through decentralisation and judicial challenges to emergent policies are examples of such measures which stemmed from the necessary response to state failure. The restructuring of conflicting bureaucratic departments which dealt with ecological matters was backed up by an agenda of concerted pressure groups and is also central to Janicke's theory of environmental policy as an integral part of capacity building. The onset of the 'communications age' driven by information technology is seen as a source of further potential for environmental capacity as yet neglected (*ibid.*). It is here as well as in improved participation and bureaucratic integration that the potential for successful environmental policy implementation is seen, in both short and long term (*ibid.*).

Regulatory Discourse Theory

The Brundtland Report of 1987 was the catalyst for introducing the concept of sustainable development and was successful, despite the difficulties presented by the fractious nature of this form of social planning. In a parallel with Janicke, Hajer outlines environmental policy regulation as a central feature of the current ecological debate. This policy is seen as the institutional bureaucracy of the state with social constructivism and discourse analysis replacing the direct activism of the 1970s radical environmentalism. These changes reflect the acceptance of a policy-led approach to implementing radical reforms, that is to say, activism from within rather than protest on the periphery of society. Within this conceptualisation is the thought that institutional arrangements must be both proactive and reflexive, the results of which are tested in the public domain through debate and investigation. The dominant paradigm in this shift in the ecological debate is EM. In the 1990s, in relation to policy and institutional change ecological planning became an aspect of the response of the member states of the OECD to the crisis of legitimacy and state regulatory failure. This has led in part to the anticipatory future vision which is inherent in ecological policy playing a part in the creation of overall policy by first predicting the most likely conflicts and their resultant scenarios (*ibid.*). The links between technological and scientific innovation and the flexible nature of ecological policy responses are examples of how current and future discourse between institutions, government agencies and ecologists will shape ongoing environmental regulations. They are linked by a common origin of necessity and, together with an outward looking dynamic, focused on the future. These ongoing impulses are rejuvenated by a continuing reflexivity that provides the opportunity for ecological policy to take on a greater significance in the face of rapid technological changes and their spin-off ecological impact. This can be achieved by an analysis of social forces or by 'providing the sign-post for action within these institutional practices' (Hajer, 1995, p. 264).

It is within this sphere that we can identify the trends that led to a transformation of the environmental situation leading to EM's position as a policy determinant of once unbridgeable forces. This acceptance of a 'plurality of discourses' is reflected in EM's emphasis on open government with integrated and 'horizontal' levels of dialogue between various agencies. 'Policy discourse' revolves around differing capacities within social structures which have influence on the shift towards new criteria such as anticipatory policy making and an emphasis on the fiscal lengths of environmentalism. Central to this is the concept that nature is a resource in itself, necessary to the public good, making it difficult for polluters to escape their responsibilities by a public now aware of its right to a clean environment. Hajer sees a major role for environmental groups in future policy making as an active third party between institutions and concerned civic groups. It would act in parallel with these actors, leave early environmental radicalism behind and move to what he describes as 'reconsideration of existing participatory practices ... based on the existence of a comprehensive environmental problem. EM acknowledges new actors ... and shows itself in an opening up of the existing policy making practices' (Hajer, 1995, pp. 28, 29).

In acknowledging EM's economic benefits and in emphasising the cost efficiency of environmental approaches ecological policy holds out the promise of an effective method of capturing the imagination of both institutional and corporate policy makers. This will place the ecological agenda in the corridors of power as well as the capitalist boardroom in a way which would have been inconceivable from previous green movement approaches. However, the pro-business acceptance of increased sustainable growth can put ecological policy makers at odds with 'deep green' environmental activists. Many 'deep green' activists feel that EM is a ploy to diffuse widespread ideological protest. It is a claim to which environmental policy lies open along with its acceptance of aspects of modern technological advances. Ultimately, it remains to be seen as to how effective any ecologically driven policy will be, bearing in mind the ever increasing capacity for global environmental disaster and the proliferation of nuclear arms in states with little or no political stability. Conversely, it is this very instability which places the potential of any ecological policy at the centre of Western democracies' attempts to address both environmental and cultural questions.

Albert Weale: New Political Approaches

Weale's theories on EM echo those of Hajer, particularly in relation to the role of creating new approaches to policy matters and in the resolution of democratic difficulties at both local and national levels. For Weale EM is an ideology within itself, able to draw on large-scale support within the political spectrum. It is on this matter that a slight divergence of theory can be established between Weale and both Hajer and Janicke. Clearly, EM represents a new opportunity in the realm of diminishing social movements and could address the crisis of legitimacy being faced by all strata of the political spectrum. At the cusp of this ideal is the need to maintain a strong environmental ethic at the centre of the bureaucratic mechanism which drives policy, creating an ecological dynamic on two fronts at both institutional and activist levels. It is this distinctive feature that provides the theoretical conceptualisation of an overriding EM belief system within any policy determinants.

Weale highlighted these theories in his book *The Politics of Pollution* (Weale, 1992). He acknowledges the distinction between civic activism and participation and regulatory bureaucracy which is a feature of pollution control policies in Western governments. In a parallel with both Hajer and Janicke, Weale sees the need to close the gap in relation to these strands of discourse within the ecological debate. He establishes the 'post-welfare state sector' (*ibid.*) nature of pollution controls, outside of traditional public spending. He outlines how the potential for disharmony between the market driven industrial and government agencies on the one hand and environmental policy makers on the other is avoided by the use of central policy initiatives such as taxes on polluters or even educational drives that effectively reclaim the element of control in the environmental debate. This in turn gives the appearance of a strong ecologically friendly system, led from above, which leaves no appearance of inter-departmental discord that could be seized upon as evidence of inaction by environmental radicals. There is, as a result, an ethic of professionalism in the field of environmental policy making which creates strong links with the scientific and technological experts, groups that are not noted for their radicalism. These interrelated factions may be a major determinant in the eventual creation of any regional environmental policy and ultimately shape that policy in such a way that original ecological theory cannot always anticipate.

Essentially there is a link between economic growth and the need to build environmental policy around increasing consumerism in Western society. In acknowledging this fact we see the advances that can be made from environmental policy which presents pollution control as a consumerist lifestyle choice, or in the extreme, as a luxury item which must be aspired to. By raising this consumerist ante the risks of alienating poorer nations must be acknowledged, but the transglobal nature of pollution that often results in industrial nations relocating their heaviest polluting industries to third world and developing countries must also be accepted. These nations are often then left with a choice of alleviating want in their own countries or destroying their environments by accepting heavily polluting industries. However, the possibility of raising environmental standards to their highest capacity in the wealthiest countries, which are the largest consumers and the worst polluters, exists. This is a departure from Janicke's notion that high standards of living and stable, liberal governments create the best potential for being of benefit if the global nature of the ecological crisis that reaches beyond the frontiers of the individual nation state is acknowledged.

The most important aspect of this concept is the need for technology to develop in a far-sighted and efficient way, ever conscious of the environmental impact its expansion has and will have in the future. Without this, Hajer sees merit in radical ecologists who question what Timothy O'Riordan called 'certain features of every aspect of the so called Western democratic (capitalist) culture – its motives, its aspirations, its institutions and performance and some of its achievements' (O'Riordan, 1981). This places Weale's view of eEM on a very fine tightrope with radical analysis being used to reinforce the ecologists side of an EM/sustainable development 'deal'; in effect, going forward with a policy paper in one hand and a 'monkey-wrench' in the other. Any environmental discourse remains open to the varying social forces which are part of a complex interplay between competing analyses, both green and technocratic. However, Weale remains optimistic in the outcome of this interplay due in part to what he describes as the 'socio-cognitive dynamics of the discourse-coalition that shaped up around eco-modernist story lines' (Weale, 1992). In other words, by establishing an eco-modernist discourse in the heart of the policy debate environmental theorists have influenced the language of future debates creating an element of perpetuation in the way EM holds sway as an established policy parameter. This ultimately results in a tendency towards more eco-sensitive and anticipatory implementation of policies and development overall.

Risk Society

Writing in *Politics in an Age of Risk* Beck outlined some of the shifts in industrial society from a sustained productivity to the need of late, fragmented capitalism to dump dangerous industries and pollutants on developing nations in order to avoid Western consumerist protest and substantial costs in both fines and restoration of damaged sites (Beck, 1995). Beck defines this transglobal 'dumping' ethic as a factor of 'the Risk Society' which created a divergence from the moral expectations implicit in such organisations as the United Nations (UN) choosing instead to reject established means of political interaction. This is linked closely with problems of regulatory failure and the crisis of legitimacy faced by Western politics as free-market institutions and media groups move into the political void, therefore, leading to a globalisation of environmental problems as precious structures set in place by the UN or other treaties become threatened by a 'profit at all costs' rationale.

Beck outlines the need for a political rejuvenation in the face of both new technological challenges and the growing tendency to favour the immediate needs over any ethical or visionary idea of social and global requirements. Science and technology have fuelled this process and buoyed up by the weakened state of institutional policy making have seen their advances tested and introduced with a decline in the regulatory tests or inspections. This led to an increase in transnational pollution together with the establishment of many environmentally damaging industries in developing nations. In his article Risk Society and the Provident State Beck details the 'risks created by the momentum of innovation' (Beck, 1996) which seems to have gripped Western Society. He sees the shift from 'Industrial' to 'Risk' societies occurring when hazardous waste becomes uncontrollable and the dangers posed by this hazardous pollution becomes the centre of both public and private debate. Institutions still go through the motions of appearing to be in control of these situations but the influence of fear caused by the increased risk has come to dominate the mechanisms of society. Irresponsible and hazardous production has led to a crisis in the agencies charged with dealing with this explicit threat. Beck established the main reasons for this situation as stemming from a growth in the individualisation of society, the crisis of political legitimacy and the influence of technology on the environmental debate, part of what he sees as the flawed nature of EM.

Beck goes further in his condemnation of 'the automatic operation of autonomous modernisation processes which are blind and deaf to consequence and dangers'. His fears for this form of 'reflective modernisation' are based on society's inability to deal with risks it introduces. He goes on to describe 'conflicts of accountability' (*ibid.*) which arise from the consequences of hazardous overproduction. 'These consequences are worth noting and include large-scale nuclear and chemical technology, genetic engineering, threats to the environment, the arms build up and the increasing impoverishment of humanity living outside the Western industrial society' (Beck, 1996). There is a need for social change in relation to many emerging factors including how industrial society deals in the future with nature and indigenous cultures. Acknowledgement of the dangers of continuing hazardous overproduction which poses a threat to society and the environment in new and incomprehensible ways needs to be part of this social change. A restructuring of society, with a de-emphasising of the individualist basis of that society augmented by Western socio-political concepts such as the welfare state, educational, labour and legal systems which are centred round the rights of the individual should be part of social change.

The lack of any acknowledgement of the scale of risk to society inherent in any ongoing industrialisation and its corresponding institutional decline are central to Beck's critique of both 'reflexive' and EM. Here Beck ultimately rejects the discourse between technology, politics and environmentalism presented by Janicke, Weale and Hajer on the basis that EM is modernism but it has overproduction and risk at its core.

Beck fears that EM will be another factor in the manufacturing of an acceptable image of what has become an experiment with the planet and its inhabitants in the name of profit and technological expansion. Furthermore, Beck sees a future of risks expounded by the attempts to deal with the dangers for society inherent in these risks with further technologies being engaged to confront risks presented by previous technologies; in effect, an endless cycle of ever increasing destruction in the name of progress and modernity. While these themes present a provocative and justifiable critique of the modernist ethic of linear expansion the alternative is never clearly spelt out. Beck's view could be taken as a worst-case scenario that presents a series of 'no win' analyses without any pragmatic or attainable subtexts. The discourse of EM with technological expansionism does, however, provide society with a detailed and attainable success in the policy field. This is needed in the immediate political future if environmental concerns are to maintain a foothold in the shifting political and societal arenas.

Conclusion

The background to EM was examined through the writings of Janicke, Hajer and Weale. The links between the development of an EM ethic and institutional concerns about sustainable development, the crisis of legitimacy facing Western politics and the need to address bureaucratic regulatory failure provide the backdrop to most of this debate. Janicke's arguments on the need for stable democratic structures have been outlined.

Among the criteria listed he cites the need for pluralist, democratic governments with an activist ethic in civil and voluntary organisations to create the conditions necessary for the fundamental changes needed for a successful implementation of environmental policy. This is vital for Janicke if questions of legitimacy in democratic institutions are to be addressed.

The concepts of Hajer and Weale were explored to show the differing approaches these writers take to EM. These differences centre on Hajer's regulatory approach to policy creation and implementation which contrasts with Weale's view of EM's ideological aspects. Essentially, Hajer promotes EM as a new form of political discourse which can facilitate a combination of sustainable development and controlled economic growth by presenting the benefits of environmental efficiency to the corporate sector. Taking his cue from the Brundtland Report Hajer outlines the economic benefits of policy creation through a corporate-style approach bringing together government agencies, ecological NGOs and civic groups to work in tandem with the corporate industrialists who create much of society's pollution.

Ultimately, this differs sharply with Weale's view of ecological modernisation's ideological potential as part of what he envisages as a 'social learning' capacity at social and cultural levels. It also differs with Hajer's theory in relation to a shift in emphasis away from corporate and economic concerns placing instead a strong significance on the potential benefits of a process of re-education in both public and institutional arenas in regard to environmental matters. This takes into account new social dynamics and political movements such as green politics and ecological pressure groups. Weale sees a need for the interaction of these diverse strands to help shape future educational and policy minded approaches which he feels will need to confront the corporate sector rather than working in tandem with them – a form of acquiescence which Hajer explores. Weale wishes to 'turn governments green' by making environmentalism the new focal point of the policy process as opposed to its current place on the periphery. He sees the potential for this to be achieved by integrating the public participatory sector with a revamped political structure and both sectors taking an environmental ethic as their one determinant.

In response to these views Beck highlights the risks inherent in any eco-modernist approach. He sees this as a symptom of the 'risk society' where dangerous consequences are excused and reaffirmed by maintaining a technologically dominant approach to global structures. Beck details his theory of the 'risk society' as that which allows ongoing destruction and pollution to continue as society loses sight of basic ethical approaches to productivity dumping hazardous industries and the resulting pollution on developing nations. By allowing expansionist industries to use EM as an 'eco-friendly' front, Beck feels that modern environmentalist discourses could be doing great harm to future generations. Beck rejects modernisation out of hand from its Enlightenment roots to the current global environmental crisis. He sees the ultimate challenge facing society now as being based around the re-negotiation of social groups with environmental considerations leading the way for the future.

ENVIRONMENTAL POLICY IMPLEMENTATION

Introduction

The ongoing shifts in environmental policy throughout Europe have been influenced to a large extent by EM in its many aspects. The varieties of interpretations which have been constructed around EM are reflected in the fragmentary nature of current European policy responses. The indigenous problems of each member state, in turn, make any overall community policy in relation to the environment problematic and fraught with regional tensions. In the article 'The Implementation and Enforcement of European Community Environment Legislation', Collins and Earnshaw (1992) point to the void between policy agreement and subsequent implementation preceding the introduction of the Single European Ace in 1987. While this Act gave certain legal guidelines regional variations have persisted.

While legislation is continually being introduced implementation remains sporadic. Collins and Earnshaw highlight some recent attempts by the Economic Community (EC) to increase policy realisation. These include the Fourth and Fifth Environmental Action Programmes which are statements in the Maastricht Treaty relating to the adoption into domestic law of EC directives and declarations such as that of the European Summit in Dublin in 1990 which states 'Community environmental legislation will only be effective if it is fully implemented by member states' (Collins & Earnshaw, 1992, p. 214). Janicke's criteria for strong democratic structures to facilitate policy implementation is echoed in Collins and Earnshaw who see the need for improved structures in the domestic managerial approaches to policy goals by member states. They outline the need for improved coordination between existing indigenous implementation bodies and any current or future EC inspectorates. However, the unrealistic short-term probability of any such development is acknowledged.

The impact of the EC legislative process, enforcement procedures and suggestions for potential improvements in this process will be detailed in this chapter. From this detail conclusions can be drawn in relation to the adaptation of indigenous policies in the composition of EC laws. Furthermore, the need for cooperation between agencies involved in policy planning and implementation is expressed with an earlier anticipation of implementation needs being highlighted. Enforcement of this policy is seen as crucial with stronger links between member states and EC authorities being called for. This includes plans for further harmonisation or 'osmosis' between national agencies and the EC authorities. Improvements suggested include changes in managerial approaches in relation to decision-making processes, the potential establishment of regulated national standards and an overall European Environment Agency to deal with inspection and enforcement. Concerns about policy implementation at a national level are a central feature of the environmental debate as is how this area has become an integral part of any overall environmental performance at national and global level. Particular attention is paid to questions arising out of economic and fiscal difficulties and the bureaucratic tensions that can result. The national cases of Holland and the United Kingdom are presented to highlight these issues, based on responses to ecological challenges through the 1960s to the present. The next four sections will each explore an area of national policy which has influenced Irish environmental approaches. The first section outlines the role of European Union (EU) integrated policy and its implications. The second looks at the specific problems of the NEEP in the Netherlands. The third explores British policy styles in relation to ecological matters. The fourth and final section looks at the area of cost efficiency and economic criteria surrounding environmental policy implementation.

One of the major causes of concern for environmental policy makers is the conflict between economic interests and the inherent costs in maintaining natural resources through pollution levies or market constraints. This becomes most evident when the basic equation of jobs equalling votes becomes the prevailing criteria for any policy maker as any disruption of market forces by the state is usually unwelcome and rigorously resisted by industry. The question of environmental regulation in the ideological base of the national and European political spectrum then arises. Despite the rise of Green parties throughout Europe a marginalisation of environmental issues that challenge environmental policy exists. The fact remains that to economists and industrialists natural resources are simply underdeveloped and underutilised commodities waiting to be exploited for commercial gain. In recognising this it becomes clear that many obstacles are placed before policy makers dealing with environmental issues even prior to any coordinated position being agreed. Politicians could improve this situation by recognising the electoral benefits that would result in an investment in ecologically friendly industries and improving environmental practices at domestic level. Here the difficulties of policy such as coordination, integration and regulatory behaviour can be confronted. The difficulties encountered by proponents of EM in the face of active policy decisions both nationally and in the EU as presented by Sbragia will be examined (Sbragia, 1996). An analysis of Dutch environmental policy cases with references to the shifts from a compartmentalised, bureaucratic approach to a more holistic dynamic will be presented.

In the main, this shift reflected a policy change from 'end of pipe' to more preventative measures reflecting the oscillating perspectives of both policy makers and industrialists to transformations in technology. The second part of this analysis highlights the problems faced in the 1970s and 1980s due to the lack of a coordinated policy both regionally and in the wider European community. Specifically, we will deal with the difficulties faced by the state in relation to both economic growth and its subsequent environmental problems and how different forms of pollution affect state policy. There are questions related to problems of substantive legitimacy and political accountability in relation to the debate between democratic majoritarianism versus 'zero-sum game' or 'win-win' policies which have been a feature of the holistic policy response to the concept of EM. Writing in *Regulating Europe* Majone highlights the introduction of new efficiency standards as being central to a successful policy implementation (Majone, 1997). However, he also accepts the limitations of regulatory directives which are used to restrict policy parameters rather than facilitate substantive changes in overall approaches. Majone points to the importance of 'value creation' and 'value claiming' (*ibid.*) to the introduction of efficiency standards which are necessary to integrate the policies of EU member states. This integration process is of particular importance to environmental policy makes, as established by Michael Skou and Liefferink (1992). However, at this stage a fuller explanation of environmental policy as integrated in the EU would be appropriate.

Policy and Integration in the EU

It is argued that governments have allowed the EU to dominate environmental policy due to the difficulties faced with industrial regulations and ongoing shifts in technology that make implementation expensive and therefore politically volatile. Because of this policy is now driven by what are described as 'Leader States' such as Germany, Denmark and Holland.

The remaining states of the EU are labelled 'Laggard States'. Policy implementation is compromised by the conflict of interests that results from this dichotomy. Ultimately, any environmental initiative by these 'Leader States' is delayed and diluted by the EU in order to facilitate overall integration of policy. This was described by Haas as 'the Leader-Laggard dynamic'. The worst 'Laggard' states include Ireland, Greece, Italy, Spain and Belgium which are also the least economically developed states of the EU. While Greece, Italy and Spain have historically weak democratic structures pointing to difficulties on the policy level placing environmental policy, with its accompanying taxes and restraints, at a particular disadvantage.

An additional problem is identified when inter-EU competitiveness is examined. As industrial policy remains unregulated between member states any subsequent environmental policy is usually the result of advances made by 'Leader' states. But more often the advances stem from initiatives made by German policy makers. Indeed, Sbragia identifies Germany as the 'key leader' in this regard and in certain cases, such as water policy, existing German legislation has been co-opted into overall EU policy directives and laws which reveals a startling democratic deficit in regard to internal environmental policy in member states. German environmental policy was developed in the 1970s and 1980s due to acid rain problems and the need for legislation for large-scale industries which were often major polluters. When international organisations such as the Organisation for Economic Cooperation and Development (OECD) or the European Union wanted to draft environmental legislation, German policy was the strongest existing such legislation to draw on.

This led to a range of environmental issues being addressed by specific German cases which may prove problematic for certain regional variants due to the contrasting degree of industrialisation in different states. According to Weale 'the political agenda of Germany has been decisively changed' this was due to environmental issues (Weale, 1992). While problems of competitiveness remain Germany's high economic standard had already been achieved in contrast to many other EU member states. It can even be said that advantages were gained by German industries due to an ability to anticipate EU directives based on existing domestic legislation. This was reinforced as EM became a prominent feature of German policy making and in turn environmental aspects became central to economic forecasting. The advantages can be seen in 'the principle of precaution' which is now central to both German and overall EU environmental policy highlighting the economic benefits of ecological concerns and pollution controls. These policies remain the standard to be met as part of member-state deregulation and harmonisation of environmental criterion.

Problems of Coordination in the Dutch Case

The research of Van der Straaten includes the results of the introduction in 1989 of the Dutch National Environmental Policy Plan (NEPP). While the NEPP is not without its critics it can be seen as an important example of both the problems and the potential of EM when implemented as actual policy. As the Dutch government began to respond to the Netherlands's heavy pollution in the 1980s following a decade of environmental debate it became clear that new institutional structures would be needed to successfully implement any EM policy. A new cabinet ministry was set up which would involve planning and policy coordination with industry. The then liberal government linked its environmental policy with its plans for industrial deregulation; in other words, industry was to police itself for pollution levels based on government set efficiency standards. This policy included a new decompartmentalised approach reducing bureaucratic red tape and integrating the various government agencies responsible for the environment and planning.

These changes were introduced after decades of problems with uncoordinated policy plans that left environmental issues on the periphery of government policy, outside the remit of any industrial or planning agenda. In the 1970s economic considerations far outstripped any environmental concerns which were at this stage confined to a few academics, and like most of Europe and the United States, part of the prevalent counterculture. However, a growing eco-awareness had its roots in fears about acid rain around the Rhine Delta. In 1972 a government Framework Policy introduced an emission ceiling of 500 m kg (Hajer, 1995). Prior to this the 'polluter pays' principle was central to Dutch government policy with levies and subsidies introduced by the Ministry for Environment on its establishment in 1971. The problems facing the three distinct areas of Dutch municipal government in relation to the environment were soon apparent; the layers of bureaucracy involved prevented any comprehensive policy on specific issues being implemented and maintained. An additional problem in the Dutch case was the traditional open politics of the country; with coalitions the norm policy is always faced with compromise. This gave rise to an economist led overview taking prevalence in regard to environmental policy making more incisive procedures unlikely.

By the 1980s national policy was being influenced by EEC directives adding another level of bureaucracy to the system. In addition to this a culture of compliance existed between successive governments and indigenous industries in avoiding certain EU directives for economic benefit. Lubber's neo-Liberal led government wanted to reduce state involvement in Business affairs and withdrew many of the procedures which had been introduced to maintain certain environmental parameters. EM had introduced the concept of environmental efficiency to the policy process; this was now turned on its head as 'eco-modernist' ideas about the deficiency of the legal regulatory approach and the advocacy of ideas about internalising environmental care had a strong affinity with the neo-liberal goal of deregulation' (Hajer, 1995). However, by the end of the decade problems such as high acid rain levels left government with no option other than a renewed policy programme in regard to the environment. It had become clear that the old Environment Ministry with concern for planning and housing had no real teeth when confronted with economic factors either at cabinet level or in the marketplace. A major catalyst in this re-think was the Brundtland Report of the WECD in 1987 which proposed the introduction of guidelines that promoted sustainable development policy. This ultimately led to the Dutch National Environmental Policy Plan (NEPP) in 1989. The main thrust of the NEPP involved a detailed analysis of what were then the most immediate environmental problems facing the Netherlands, including:

1. reducing substance emissions by 70 to 90%
2. overall energy savings
3. improvement of environmental standards in the production process
4. improved transport policies
5. reduced fertilisation in agriculture
6. intensifying long-term environmental policy

(Van der Straaten, 1992).

A major concern of the NEPP was the breaking of economic and ecological substance cycles (Van der Straaten, 1992; Weale, 1992). This view is close to eco-modernist ideas of environmental efficiency, that is to say environmental problems have severe economic and social repercussions. Links were made between renewable cycles of ecological 'raw materials' and maintaining substances for economic productivity. Furthermore, it was now becoming apparent that human interference with substance cycles was taking its toll on the global ecosystem placing conditions for human existence at any level in jeopardy.

A summary of the NEPP's environmental goals to be implemented by 2010 includes:

- responsible citizenship,
- allied responses with other nations,
- promotion of technological development towards sustainable development,
- structural source-oriented measures,
- further development of integrated policy including issues such as environment, responsibility of actors, implementation of risk liability and the utilisation of financial and fiscal instruments.

These goals were introduced to a system with a history of problems in regard to environment laws since their inception in the 1970s. There were difficulties in getting proper financial backing from the Treasury so levies and taxes were applied in a rigorous adaptation of the polluter pays principle. Pollution was to be rendered financially unviable for industry. But this in turn increased levels of bureaucracy in local and state government departments. This soon became so costly that monies gathered from levies were outstripped by bureaucratic expenses. As a result the polluter pays principle became unpopular in government circles by the late 1970s. It was replaced by a single fuel levy with responsibility for this procedure shifting from the Ministry of Environment to Treasury. It soon became clear that tensions between those Ministries, allied to ideological problems relating to tax increases for the centre right government, had the potential to curtail much of the content of the NEPP agenda.

Subsequent elections gave control to a coalition of Lubber's Christian Democrats and the Social Democratic Party who then gained the Environment seat. Following these changes the NEPP was introduced as policy with economic factors being measured against what became known as 'the ecological norm' (Van der Straaten, 1992). However, tensions remained between the Environment Ministry and the more traditional economic views of the Treasury. Without an overall change in policy by the entire government environmental policy can never be properly implemented and the results never rise above peripheral and incremental changes. Ideological questions relating to the role of the state in the marketplace and of industry's inherent obligations to society and the environment continue to cause divisions in Dutch governmental circles. While government policy is dominated by financial considerations policy advances such as the NEPP are often paralysed. This leaves the role of environmental NGOs in a position where they must continue to put pressure on government agencies while simultaneously keeping the environmental agenda alive in the public domain.

British Policy Styles

British policy styles were more influenced by domestic UK agendas than continental European policy which has an EU led dynamic underpinning much of its legislation concerning the environment. The distinctive nature of British policy styles are examined now while the challenges to domestic policy presented by EU legislation and the indigenous problems which create certain tensions between domestic UK politics and British-EU policies are outlined.

While the distinctive features of UK policy based on internal factors are a particular aspect of that nation's legislative response to many issues Richardson and Jordan highlight the role of each EU member state's policy style in influencing subsequent policy positions. They see this form of internal policy style as being a distinctive facet of member state policy rather than providing each state with a clearly defined or unique national analysis of EU legislation. Essentially, they see close similarities between member state's policy styles describing them as 'convergent' while accepting that localised variations play a role in domestic responses. Consequently there is a shift in the structuring of this response ranging from the point where a potential policy is agreed to the actual implementation of a coherent policy initiative.

Within the mechanism of this implementation distinctive national characteristics become crucial to the shaping of policy and it is this, in effect, that shapes policy styles throughout Europe. As a result, dominant formulae for implementing policies become a feature of a nation's bureaucracy. However, certain anomalies remain in any standardised definition of policy styles. These can be summed up as follows: 'Even if a predominant style can be identified, exceptions will always occur ... One of the biggest difficulties in trying to identify a predominant style is, of course, present by the now familiar phenomenon of the sectorisation of policy making' (Jordan & Richardson, 1982, p. 3).

What becomes clear through this definition is the role of indigenous bureaucracy in the shaping and influencing of domestic policy. This in turn affects the implementation of any such policy giving EU legislation its localised variants while heightening inter-state tensions. Elements of this bureaucratic discord can manifest itself within the national body politic taking on ideological dimensions and increasing anti EU feeling, as has been seen in the British case. As a result of this, certain issues that are linked with the EU directives, such as environmentalism, become further isolated from local political delegates and lead to increasingly alienated responses from a public unsure of domestic political positions in relation to an apparent EU bureaucratic monolith in Brussels.

These incremental changes have lent themselves to the formation of any distinctive British policy style. As this distinctive policy becomes standardised it gains acceptance among politicians, bureaucrats and the public, ultimately providing a 'national' policy style. It then becomes the national bargaining position adopted when approaching central EU government or other member states. Jordan and Richardson identify UK policy as deriving from a 'logic of negotiation' based on what they see as five distinct aspects of British policy style (Jordan & Richardson, 1982). These aspects include sectorisation, clientelism and consultation, instituionalisation of compromise and development of exchange relationships (*ibid.*). Central to this style is a policy of ongoing accommodation through bureaucratic negotiation, with an established position being presented for acceptance by the parties involved, rather than having a potentially fractious debate in order to reach a potential outcome.

British Policy Background

Taken as a distinct entity British policy style is at its most apparent when confronted with EU legislation. While much of European policy had become binding by the late 1980s long-standing British attitudes of unease with the nature of much of this policy persisted. British policy has had an inconsistent evolution in parallel with the rise of what is seen as German led EU legislation. As such environmental policy can be seen as a good example of these underlying tensions in this regard and the history behind this dichotomy will be outlined in the next section and based, in particular, on Lowe and Ward's findings in *British Environmental Policy and Europe* 1998. These writers emphasise the fractious nature of overall EU policy, which they see as a reflection of inherent domestic cultures, and as a result they see their examination of British policy perspectives as only one of many pan-European distinctions rather than highlighting British tensions as an outstanding area of policy difficulty.

In addressing this issue the role of distinctive national policy is detailed as being 'the basic unit of the EU ... having particular conditions which shape policy positions ... EU policy making is an amalgam of national agendas' and that 'national governments hold the key positions in shaping European policy through the Council of Ministers and ... domestic policy' (Lowe & Ward, 1998). Essentially, it can be seen that domestic agendas with their own regional economic and social circumstances take precedence over any EU considerations and in any such case EU issues can be relegated to a secondary position to suit domestic policy. What becomes evident is that resulting tensions increase as EU directives become the policy norm in individual member states often as the result of an absence of domestic action within these states. Britain's tradition of 'Euro scepticism' has led to many difficulties regarding that nation's dealing with certain EU directives, environmentalism and 'the principle of precaution' being a case in point.

Lowe and Ward identify the historical backdrop to a 'semi-detachment' to the EU power base in Brussels despite what they see as 'any realistic alternatives' being evident. This history has its roots in traditional elements of an 'island mentality' heightened by a strong imperialist tradition with worldwide trading links, the ongoing relationship with the United States and the Dominions and a reluctance to accept a modern role as 'a middle ranking European power' (*ibid.*). This shift can be contrasted with the traditionally strong role of domestic British environmentalism specifically in areas such as wildlife conservation. International environmental groups such as the World Wildlife Fund (WWF), Friends of the Earth (FOE) and the Institute for Environment and Development (IED) were then established and these organisations had a major impact on the global ecological consciousness in the 1970s.

Another, traditional influence on British environmental policy was the ownership of large tracts of lands by the establishment which led to a divide between common land and land restricted to use by the upper classes. Codes of law in relation to rights of passage, wildlife management, hunting and fishing practices were developed by the upper class which created that divide. Consequently, an insular perspective developed in relation to the onset of EU environmental directives. Furthermore, EU legislation tended to draw influence from German and Dutch regulatory experiences without references to long-established British ecological practices and this heightened the rising resentment to Brussels.

As a result the emphasis has been put on mid European ecological viewpoints leaving Britain exposed as 'the Dirty Man of Europe' (Lowe & Ward, 1998), a label that can only be seen as having a prejudice in favour of European Greens. This view of Britain as falling behind European standards was taken up b the British Green Movement in its opposition to Thatcher's government policies during the 1980s. A particular example of this was the Windscale/Sellafield

Nuclear Power Plant and the anti-nuclear missile base campaign at Greenham Common. Ultimately, the EU came to be seen as 'an environmental beacon' (Lowe & Ward, 1998). Domestic political agendas, Cold War alliances with the United States, cultural and historical factors mentioned, above all, lead to a dichotomy in EU and British environmental policy.

Cost and Efficiency

Skou Andersen addresses the question of economic criteria and its negative effect on environmental policy in *Governance and Green Taxes 1992*. In his examination he explores the effects of eco-modernist tax and levy systems on indigenous industries and details the difficulties faced by legislators and governments. The difficulties rise when administrations prefix economic measures to environmental efficiency standards without allowing for the inevitable market fluctuations which then expose any environmental policy to fiscal restrictions. His response is to set eco taxes within an overall policy that has been reshaped in the approach to pollution threats as opposed to guidelines set which can then be ignored or rallied against by industry at will. Pre-emptive 'green' taxes are built into the state's annual budget and would be above inter-departmental regulatory problems. The key to this approach is in the costing of each unit of pollution which anticipates market fluctuations and allows industry to make pollution control an economic target with the efficiency argument of eco modernisation (EM) at its core.

One of the difficulties with pre-set eco taxes is that the ongoing development of new technologies is left out of any ecological guidelines leaving industry free to continue to produce more inefficient and polluting products in an efficient and cost effective way. This compounds problems in pollution control as green taxes and efficiency standards mask ongoing pollution from non-conducive technologies. While technology and environmentalism remain separate it becomes impossible to formulate a comprehensive policy. High costs are often shifted into the public tax arena making pollution control unpopular and politically divisive. This affects short-term policy as well as any future initiatives leaving environmentalism marginalised. Questions of expense are passed on to the next generation who, in turn, may find ecological damage to be irreversible and economically overwhelming. Economics rarely has a future vision and the value of the environment both now and in the future remains outside the estimations of economic experts. Final criteria for environmental policy are also going to be open to influence from pressure groups with vested interests leaving any such policy with a potential democratic deficit. Ultimately, economic and technological actions are intrinsically linked and form a powerful lobby group in governmental circles. Taking this into account is a strong factor in any argument for the completely restructured approach to environmental issues at policy level, including the coordination of government agencies and a holistic, socio-economic approach.

Conclusion

The main thrust of this chapter has been centred round the problems for policy makers when confronted with the full rigours of fiscal rectitude and bureaucratic intransigence. It has highlighted how even a major policy breakthrough such as the NEPP is open to the influence of both inter-party ideological struggles and inter-departmental coordination failure. This was shown by exploring EM's influence on active policy and attempts being made at holistic approaches which only succumbed to tensions within departmental coordination and integration. These problems compounded Lubber's government's attempts to implement the NEPP. Environmental policies were diluted due to economic and electoral considerations.

The challenges of integration policy within the European Union were explored as well as how countries like Germany, which has advanced environmental laws in place, come to influence and even dictate much of EU policy. The concept of 'Leader' and 'Laggard' states illustrates this point. Problems of coordination at a European policy level such as how differing forms of pollution like acid rain influence indigenous ecological policies were outlined. This led on to an examination of the problems of policy integration in the Dutch case with a detailed analysis highlighted. In particular, Dutch reaction to the Brundtland Report and the emergence of EM as a policy tool is considered placing specific emphasis on the problems and success of the NEPP through the 1980s and 1990s. A study of the British case is also provided which examines differing policy styles.

Localised variations have given UK policy a negative tendency towards overall EU directives with very real consequences for both British environmental planning and policy implementation. This fractious relationship with the EU has taken on an ideological perspective which increases the difficulty faced by subsequent planners. Theories are presented in relation to the tensions arising from UK/EU implementation conflicts. There are problems for all member states in this regard with many of these tensions having domestic political origins. EU legislation and directives challenge and ultimately shape domestic policy styles. Member states are described as having an overall convergence in their policy styles with regional variations stemming from local political debates. The role of bureaucratic interpretations in the sectorisation of policy style based on the sum of aspects of indigenous problems is acknowledged. This provides member states with their initial bargaining positions making an acceptable vision for environmental policy on an EU-wide basis a near impossible goal in any policy area.

Following on from this the negative impact of rigid economic parameters on environmental policy as well as the need for shifts to holistic approaches are presented with an emphasis on the problems and benefits of 'Green' taxes and levies. A main concern put forward here is the difficulty faced when economic fluctuations disrupt pre-set tax and levy points leaving the way clear for industry to avoid some or all of the levies it could potentially face in regard to pollution controls. This raises questions of how environmental policy can react and interpret changes in technology and if it is possible to link technological advances with holistic environmental approaches to socio-economic planning. The results of these inquiries remain to be seen. What has become clear, however, is the need for concern with such planning by all including industrialists, lobby groups and policy makers to consider the costs involved in continued environmental damage both in fiscal and planetary terms.

IRISH ENVIRONMENTAL POLICY

Introduction

Ireland's history is dominated by its population's relationship with the land. A strong social and cultural tradition has grown around this relationship and the resulting network of pastoral, agrarian conservatism has led to a culture of clientelism and corporatism that makes new policy approaches difficult to implement. Many of the responses to the challenges of the environment in Ireland are led from the outside, particularly in the context of EU directives. Domestic political responses, such as the introduction of the Environmental Protection Agency (EPA) have had a mixed reaction while the political culture of the country has prevented a strong ideological campaign from gaining influence. We will now look at the background to the introduction of the EPA and examine whether its functions have improved environmental capacity or is it being used by sectional interests to enhance industrial opportunities in the State. Then the position of the Irish

Green Party and their role in the Irish political arena is examined. Following on from this the civic responses to environmental matters will be analysed outlining ongoing problems with equality of public access, difficulties with legal structures and local authorities. The Conclusion explores the concept of a democratically guided democracy which places environmental concerns at the core of State requirements.

The Emergence of the EPA

What becomes evident in any study of environmental policy is that the significance of EU policy in the development of environmental procedures throughout Europe cannot be underestimated. Specifically German, Dutch and British case histories must be examined in order to establish the sources and motivations behind any eventual EU policies. Over the previous decades the impact of a heightened ecological focus has brought these debates to the fore in Irish life as a previously agrarian society industrialised and the Celtic Tiger economy gained strength. High profile cases such as Mullaghmore and the Mutton Island Sewage Treatment controversy heralded the arrival of many diverse groups such as the Burren Action Group and the Save Galway Bay Campaign. These groups seem to be the result of spontaneous community protest rather than stemming from any history of ecological activism in the country. However, the Irish Green Party has representation at local, central and EU levels and eco-protesters have come in and out of the political picture since the Carnsore Point anti-nuclear protests of the late 1970s. That protest was the result of action taken by Greenpeace and the 'No Nukes' campaign and was influenced by its larger British counterparts. There are parallels with that event and the eco-warriors of the Glens of Wicklow who took to the trees to prevent the destruction of the Glen of the Downs for a motorway development.

In effect, the Irish landscape has become a site of conflict in the ongoing struggle for economic and infrastructural development. The lack of a strong ecological dynamic in Irish politics as seen in its over reliance on British and EU policy innovations to deal specifically with Irish test cases has compounded the nature of this problem. As economic and industrial growth increases and EU directives highlight further aspects of ecological threat which need intervention, a tension has developed in Irish society that is manifest in the increasing number of eco protest groups. The traditionally weak nature of the Irish political system has thus far been unable to respond with any efficiency to the environmental debate despite the adoption by the centrist parties of token 'Green' manifesto promises. Clearly, the shifts in the politics of environmentalism have not been fully realised by the mainstream political establishment in Ireland.

These shifts have been centred round a plethora of issues that have their roots in the ecology versus development debate. Arguments and conflicts are in the areas of planning procedures for infrastructural development, large urban housing estates with their road and sewage offshoots, IDA sponsored industries polluting the waterways, agri-industrial spillages destroying lakes and fish stocks and the hidden pollution of companies such as Syntex in Clarecastle or the location of major incinerators or 'super dumps' in rural areas, and have taken precedence in recent years. The debate about the introduction of genetically modified (GM) 'Frankenstein foods' into the marketplace by supermarket chains has the headlines. While the public at large seems to have little desire for such products manufacturers like Monsanto continue to supply their GM goods to the market. Out of this came the group *Genetic Concern*.

While these pressure groups represent a public response to environmental concerns government agencies have a poor record of environmental action. Environmental policy has been largely in the hands of Irish regional and local authorities. Control over ecological issues was consolidated in 1977 by two pieces of legislation – the Local Government (Water Pollution) Act and the Local Government (Planning and Development) Act. The legislation gave local authorities primary responsibility for air and water pollution controls as well as sanitation and water management. However these Acts reflected a need to implement EU directives rather than representing any new ecological awareness on the part of government agencies. They also had the unfortunate result of establishing local government as a conduit for agrarian and corporate interests which aimed to minimise the affects of environmental legislation on their inherent production costs.

As a result of pressures to increase employment rates the 1970s and 1980s were marked as an era when environmental concerns wee often relegated in favour of attracting inward investment and multinational jobs. Many of the multinationals such as Merril Dow in Cork were grant aided by the IDA and attracted by the prospect of a less than stringent regime of environmental legislation. This placed the local authorities in a position where, in an atmosphere of near desperation for any form of job creation, pollution controls which could hinder major industrial investment were largely sidelined. The result was an influx of chemical and pharmaceutical industries establishing themselves in Ireland with regular incidences of pollution and public outcry becoming more frequent.

Public disquiet at the relationship between the IDA and multinationals was met with a policy response by the incoming coalition government in 1989. The first Green Party TD was elected to the Dáil that year. This led to the establishment of the Environmental Protection Agency (EPA) four years later (July 1993). The delay points to the lack of preparation and direction surrounding the structures and remit of this body. The powers of the EPA were to include the provision of a legislative framework to assist local authorities with their environmental duties. The EPA was to be an independent body free from the role of dealing with multinationals on the level of attracting industries into the country. Instead the EPA took charge of licensing and regulation of industries, monitoring pollution levels and gauging the response of industries to new environmental directives.

One of the most significant aspects of the EPA's new responsibilities lay in its attempt to implement policies in line with the innovative concept of Integrated Pollution Control (IPC). The central feature of IPC is based on the establishment of an overriding system of pollution containment which links together the various features of the environment such as land and air and recognises their intertwined relationship. Scannell outlines the areas covered by the EPA's Licensing Regulations (1994) where a system of integrated pollution control would bring a diversity of categories under one central authority. These include 'Minerals and other materials, Energy, mineral Fibres and Glass, Chemicals, Food and Drink, Wood, Paper, Textiles and leather, Cement and Waste (Scannell, 1995). Clearly, the broad spectrum of Irish industry would now come under the remit of IPC regulations. Semi-state and local authority activities would also be monitored under this Act. Another important area of IPC monitoring would be the increasingly expansive pharmaceutical industry. It is interesting to note that the pharmaceutical industry, one of the main industrial pollutants whose poor environmental record was a factor in the establishment of the EPA, is only subject to IPC scrutiny 'where the numbers of employees exceeds 100 but is less than 200' (*ibid.*). This section of the Act was obviously included with the input of the IDA who hoped to continue to attract large-scale major employers such as Elan to Ireland in the future, unhindered by any ecologically minded small print.

It is here that we can establish the main weakness in EPA legislation. While a large area of the environmental landscape is now under a focused and centralised structure, the criteria which surrounds the establishment of the EPA have left their imprint on the Act's legal framework and as a result there are loopholes that can be exploited in what should be strong, legal environmental parameters to shape Ireland's future in terms of the environment. For instance, Section 52 of the EPA Act outlines how that should 'provide support and advisory services to local and public authorities in relation to the performance of their functions' (EPA Act Section 52). This section of the Act deals with the highly contentious area of environmental performance by the local authority, yet it in now way specifies how 'support and advisory services' should be provided. The controversial nature of how local authorities deal with the IDA, multinationals and the agricultural sector is not legislated for, while equally important policy formats such as integrated pollution control are introduced without a clear indication of the overall basis or direction that such IPC controls are intended to achieve.

A very important feature of IPC is the concept of 'BATNEEC' or 'Best Available Technology – Not Entailing Excessive Costs'. The EPA's own handbook on IPC states that 'The technology should be Best at preventing pollution and Available in the sense that it is procurable by the operator of the activity concerned. Technology itself includes techniques and the use of techniques such as training and maintenance. NEEC sets out the balance between environmental benefit and financial cost' (EPA Guidelines on IPC p. 3).

Coming as it did into a relative void of environmental legislation and policy planning the EPA Act should have been seen as an opportunity to introduce far-reaching and comprehensive environmental protections to the domestic political scene in such a way as to bolster already existing EC Directives. Instead, political and economic considerations led to the drafting of a watered down form of legislation full of the type of legal vagaries that can be fully exploited by corporate lawyers on behalf of their multinational clients. This has implications for the Irish government and can only be addressed by the drafting of a more comprehensive body of legislation which takes on board both policy advances, environmental considerations and relegates economic factors. Any form of environmental legislation drawn up with extensive remit, such as those of the EPA Act, should have extensive and detailed powers with environmental protection and the encouragement of ecologically minded structures being incorporated as an integral part of future planning at its core.

The Irish Greens: Ineffectual Posturing or Eco-Democrats?

With an absence of any real environmental planning from Dublin and faced with a less than decisive body of legislation from the EC, future Irish governments will have an opportunity to introduce a positive and visionary environmental agenda to the domestic and international body politic. Ireland still retains a less than deserved image as a place of unspoilt natural landscapes and still developing industrial and economic infrastructure. By introducing the holistic structures that will produce an environmentally minded society in future years Ireland could provide a model of how successful economic management can be achieved by working on all the different levels of policy which have an environmental dimension and by bringing consensus to planning and industry in relation to all ecological obligations. Thus far, Irish administrations have relied on EU and British innovation to apply any form of environmental legislation so the likelihood of a major ecological breakthrough on the domestic political scene remains very unlikely. However, a far-reaching environmental agenda remains as the best political strategy yet to be fully exploited by the Irish political parties. A future advance by the Irish Green Party which has consolidated itself on the political scene is possible particularly in light of the convergence of the major parties towards the centre.

The fact remains that the Irish Greens have jettisoned their activist wing as far back as the late 1980s. In a parallel with left wing parties around Europe activist and 'militant' groups were seen as a threat from within by party leadership in advance of an overall shift to the centre. While this shift came as world politics was in the grip of a right wing political wave it has left a gap in political activism for Green parties throughout Europe. While centrist politics has grappled with 'Green' issues since the 1992 World Summit Green parties have not made the major advances that may have been possible at that time due, in part, to a disillusionment with the rigidity of mainstream politics which left a chasm between ecological movements such as Greenpeace and Friends of the Earth (FOE) and European Green Parties. This gap has equally affected the growth of the Irish Green Movement and can be seen in the distance between the position of Eco-Warriors in the Wicklow Glens and attempts by Irish Greens to attract votes from the professional classes and young urbanites. The Greens participation in the coalition government of 2007–2010 effectively killed off the party as an electoral entity. It will be a long-term project to get a Green Party back to anything like its governmental heyday, which even then never went much above 5% in the opinion polls.

Equally significant is the fact that the Irish Greens do far better in European elections than in Dáil and local elections. Mullally has outlined the Irish Green Party's shift from a 'party of protest' towards being a group that has had 'more pronounced success on the European level' (Mullally, 1997). The Irish Greens, who had been known at their inception as The Ecology Party of Ireland or 'Comhaontas Glás' had gone through a process of splits and realignments as had other European Green Movements. In adopting the standard 'Green Party' title in 1986, while retaining the Irish language suffix of Comhaontas Glás, the Party 'signalled a commitment to electoral politics' (*ibid.*).

Initial gains by the Irish Greens were soon countered by a process of adaptation of Green issues by the major centrist Parties. The business oriented thrust of the EPA Acts drawn up the Progressive Democratic Party is an indicator of how Green issues were co-opted and manipulated by government parties in order to facilitate job development. Despite this, successes for the Greens in European elections have given the Irish Greens an established position on the Irish political scene. Other successful, local campaigns have pushed the Green agenda into national politics.

These advances are tempered by the stilted nature of Irish politics where shifts in European policies are adopted in a gradual manner while local issues and agendas take priority. All major political parties have converged towards the centre and have adopted pragmatic, 'catch-all' policies, including whatever is seen as relevant in regard to the environment. In light of this, it is perhaps unusual that the Irish Greens have also staked a claim for the suburban and middle class voter. However, this can be seen as part of an overall European trend away from 'survivalist' agendas and a move towards adopting a 'green lifestyle' centred on middle class, suburban concerns such as recycling, retaining green areas like forests and the spread of genetically modified foods. Like their European counterparts Irish Greens have adopted many issues to suit the increasing trend towards suburbanisation rather than advocating any real alternative.

This has left some theorists questioning whether the Greens are as yet ready for Government and how much of a truly 'Green' agenda they can maintain in the pursuit of this aim. The question remains open as to whether ultimate goals of environmental communion, species cohabitation and universal non-violence are attainable through the political structures of Western Liberal Democracies and if not how can realistic alternative structures be realised. In this respect, Irish Green Party policies such as the Mayoral Commission on Cycling and the court case taken to force equal funding for the divorce referendum in 1997 can be seen as campaigns geared as much for individual re-election as that of a fully structured and far-reaching plan for an environmental future. It would, for instance, be interesting to see how a Green Minister for Justice in a future 'Rainbow Coalition' government would deal with an event like the 'eco-warrior' protest at the Glen of the Downs. Power provides its own dilemmas yet the Irish Greens like all other Green Parties must pursue power through the political channels which currently exist.

In addition, Green Party advocation campaigns of urban recycling and consumer consciousness rely on a high degree of civic responsiveness and responsibility. Any major challenges to the existing status quo cannot therefore be too extreme or rooted in civil disobedience as this would challenge the structures needed to highlight the domestic Green agenda. That position leaves legal action as the only recourse left to Greens who wish to challenge existing structures. However, there are problems remaining. This approach 'fails to acknowledge the importance of participation in creating a consensus to underpin the legitimacy of the regulatory framework' (Taylor, 1998). In other words, an over reliance on legal actions and challenges to planning permissions while maintaining a legislative presence and 'watchdog' image for the Greens can ultimately remove them from the civic participatory groups which are needed to back up any advances on the ground.

Because of the problems a breakdown in communications between the bureaucratic central Party structures of the Greens and their activist support base could happen that would lead to a further distancing between the general public and the party as a whole. To avoid this, the Irish Green Party must maintain an activist base with moves towards a sufficiently politicised agenda outside the increasing trend towards challenges to the legal framework of the planning process. Older ecological issues based on conservation could be revisited rather than exclusively working towards a 'greening' of suburban lifestyle. There is also a need to bridge the gap between forms of public protest which verge on civil disobedience and a plan of action that alienates both activists and the wider public through excessive legal parameters. Taylor has highlighted this need to make the channels of protest more available to the public. Open public hearings are one way to highlight and facilitate public debate over environmental issues. However, Taylor points to the weakness of EPA legislation in this respect citing how 'the agency shall have absolute discretion to hold an oral hearing which shall be conducted by a person appointed by the agency (from EPA Act 1992, p. 64).

Civic Responses

Problems of equality of public access in terms of format and agenda as well as difficulties for the public in approaching a legalistic framework are areas that are left unanswered by the EPA legislation. The lack of answers to these problems points to the structure of the Act having been built around a framework which is devised to suit the legislature with no cognisance of open public inquiry into the process being allowed for. Taylor's article highlights the major difference between this approach to public interjection or scrutiny and that of license applications by companies who are told that the EPA will go 'as far as possible to assist industry' and to 'provide pre-application clarification facilities where necessary' (Taylor, 1999 from EPA Act p. 5). This leaves the EPA independent of public or civic scrutiny while it is equally relatively free of Dáil interference. Yet the nature of its own guidelines leaves the EPA as a facilitator of industry in the areas of licensing and control which were initially meant to be the main focus of the Agency's agenda. Furthermore, the granting of licences to companies to operate a toxic incinerator despite wide public protest point to a worrying potential to override public opinion and environmental concerns in favour of business interests. Pollution prevention becomes, by extension, pollution management and even pollution toleration in order to facilitate industrial expansionism. In this respect, the EPA Act can be seen as a pro-industry 'Trojan horse' planted by big business advocates as opposed to being any major step forward in environmental protection.

It is fair to say that the whole thrust of the EPA Act and its ongoing implementation has moved a great deal from the original debates of sustainable development which had influenced calls for this type of legislation in the late 1980s and early 1990s. Writing about the future for the EC's Environmental Impact Assessment Programme (EIA) Fry puts forward a series of items that have become accepted as the standard components of sustainable development for which the EIA was established to oversee implementation in member states. These include 'ecosystem maintenance, pollution control, preservation of genetic diversity and a rational utilisation of resources' (Fry, 1999).

If these components are implemented by agencies such as the EPA or advocated in policy by the Irish Green Party then the projected outcome for Irish society should be built around a cohesive equality in that society which sees development maintained with an environmental balance; with a minimal encroachment on natural resources combined with efficiency in human production. The EPA has as one of its objectives the aim of achieving sustainable development yet the specific aspects of the Act make it possible for industries to avoid environmental responsibilities.

Concepts such as ecosystem maintenance and pollution controls at the source have given way to a legally aided system of deception and avoidance of environmental responsibility in Ireland. In many respects, the emergence of the 'Celtic Tiger' economy has negated the need for much of the ambivalent guidelines contained in the EPA Act. With industrialisation and economic growth forecast to increase in the near future the circumstances surrounding the pro-business/IDA leanings of the EPA Act have been removed. In the absence of stronger legislation Ireland has become increasingly known as a country that tolerates pollution in order to attract multinational enterprises. Far from enhancing sustainable development the lack of protections granted by the EPA Act, together with the poor impact of the Irish Green Party, has led to a situation where the environmental protections that should be provided for, either from the direction of domestic green politics or government agencies, are available only in a weakened form or absent altogether.

In the worst-case scenario the background which underpins much of the EPA legislation actually enables the industrial polluter while having simultaneously withdrawn many of the local authority's powers to intervene. These factors, together with the problems previously mentioned leave Ireland's environment exposed to the worst elements of modern industrial pollution such as pharmaceutical and petrochemical industries, toxic incinerators and agricultural waste. While policy innovations from Britain, Holland and Germany are eventually co-opted into overall EU directives which affect Ireland's environmental legislative impetus the process of continuing reactive rather than proactive environmental legislative policy is clearly inadequate. It remains to be seen if this 'eco democratic deficit' can be addressed by future Irish governments and in particular if a strong environmental agenda can be set by the Irish Green Party.

Conclusion

That there must be a paradigmic shift in the power relationships and infrastructure which exist between local and national authorities in relation to the tensions between economic expansion and environmental protections has been argued. In addition, the hierarchies of power which control EU directives must be replicated within the Irish domestic political arena in order to fully implement environmental protections in a comprehensive way. In order to facilitate such an outcome, a strengthening of the EPA Act, together with a concentrated policy agenda being implemented by interested parties such as the Irish Greens, must take place. Furthermore, civic groups and NGOs must help to establish public access to the judicial system while providing a backdrop of citizen activism and political participation which can be built around ecological issues. A comprehensive examination of government actors such as Lancefort, the private company set up under central government auspices, must be made so that matters become clearer to the public at large.

This would require a strengthening of the democratic system which is currently in a reverse process as can be seen in the weakening of the nation state and subsequent democratic deficiencies. Writing in *Democracy and*

Green Political Thought Christoff argues for 'ecologically guided democracies' where there is a 'reconstruction of the state enabled ecological requirements to be met nationally and internationally' (Christoff, 1996 from Doherty & DeGeus, 1996). This type of wide-ranging policy agenda must be comprised of a coordinated effort at local, national and international levels as ecological problems often move beyond national frontiers. Ireland's complex legal and political status involves levels of bureaucracy from Dublin, London, Belfast and Brussels being involved with environmental issues for the island as a unit. Robyn Eckersley makes the point that the traditional 'liberal epistemological premise' which holds that 'it is up to citizens and their elected representatives, freely to decide for themselves the meaning of public good' (Doherty & DeGeus, 1996). This leaves little room for the framing of ecologically minded legislation which places environmentalism as a priority in the shaping of future societies; Green parties or environmental issues must withstand the full rigours of public scrutiny through democratic elections. This induces subsequent ecological political movements to contest their agenda on a par with all other political lobbies including national self-interest and industrial expansionism. Green political parties and NGOs in Ireland and abroad must set their policy framework around existing political structures while highlighting the urgent nature of global domestic environmental problems ranging from global warming and Sellafield's nuclear power to localised agri-pollution such as fish kills.

The varying strata of the Irish social and political landscape underpin much of the environmental legislation and political activity. The major difficulties thrown up by the drafting of pro-industry legislation by relevant government agencies have been highlighted. But any response through the political arena is stymied by a poorly focused and centrist-led Green Party. If these two critical components of any national ecological agenda are weak the social responses of NGOs and civic bodies are, as demonstrated by Cristoff, unable to develop in an incisive and constructive manner through existing legalistic or political fora.

In the absence of environmental superstructures industry and planning regulations have developed with a deliberate pattern of avoidance in relation to environmental responsibilities such as pollution control. Rootes provides an analysis of what type of response is needed to arrest this gulf between the 'dilemmas' of 'institutionalism' that face alternative ecological groups, problems in the relationship between local environmental struggles and national (sometimes transnational) Environmental Movement Organisations (EMOs) and the problems facing local activists in regard to the 'development of a global environmental movement' (Rootes, 1999).

It is only by strengthening the existing legal and political structures open to environmentalists that progress on ecological protection can be made in Ireland. The links between disparate groups such as the Wicklow eco-warriors, the Burren Action Group and the actual Irish Green Party should be strengthened in order to enable a fuller debate on environmental issues at local, nation and ultimately international levels. The immediate priority of these groups should be based around a reversal of the EPA Act to provide legislation with real environmental protection as its basis. In addition, the wide-ranging freedoms currently open to planners and multinationals must be reversed. Any future legislation must fully deal with these components of the national environmental agenda in order to create the conditions for the type of ecologically based society that places the environment at the centre of its long-term plans rather than reducing it to a commodity to be exploited.

THE ENVIRONMENTAL PROTECTION AGENCY (EPA)

The Environmental Protection Agency (EPA) was established in 1993 by the Government Coalition of the time. The Irish EPA is an autonomous public body which was set up as a response to the external pressure emanating out of the 'top down' directives from the European Commission on the Environment and the European Courts. The remit of the EPA also included dealing with the 'bottom up' pressures exerted on the State by the many campaigns against the siting of multinational plants or infrastructure projects in the preceding decades (Leonard, 2005, 2006). The Agency came into being through Section 19 of the Environmental Protection Agency Act of 1992. Under this Act a management structure was created through an Executive Board comprising of the Director General (DG) and four other Directors (Scannell). The DG was appointed by the State while the other Directors were appointed by an independent body that includes further appointees of the State as well as senior members of the Civil Service including the Secretaries of the Departments of Government and the Department of the Environment. The independent selection committee is further comprised of senior figures from civil society that form the basis of the Irish State's neo-corporatist social partnership model. They include the Director of the Industrial Development Authority (IDA), the Secretary of the Irish Congress of Trade Unions (ICTU), the Chief Executive of the Council for the Status of Women and significantly, the Chairperson of An Taisce, the

Heritage Trust, which is an organisation that has been to the fore in Heritage and Conservation issues in Ireland.

The Director General of the EPA sits for a period of seven years while the term for the other Directors is five years. All Directors can be returned to their positions. Section 21(16) of the EPA Act also provides for the removal of the Director General if deemed necessary by the Government. However, the EPA remains independent of any of the Government Departments to whom it must report annually. Regional headquarters were established in Dublin, Cork, Kilkenny, Castlebar, Monahan, Letterkenny, Athlone and Mallow. The primary functions of the EPA involve the licensing, regulation, planning and monitoring of issues of environmental concern in Ireland. The Agency also contributes towards the furthering of academic knowledge through research in the field of environmental studies.

Licensing

The EPA has responsibility for overseeing the licensing of practices that may have significant polluting potential. Under 'General Functions (Section 52) of the Environmental Protection Agency Act 1992' licensing is deemed part of the overall control of activities for the purposes of environmental protection. Licences are issued under 'Best Practice' guidelines which incorporate a range of approaches including the below measures.

The prevention or elimination of pollution at all levels of production and disposal with the incorporation of recycling, reduction and reuse and the avoidance of dangerous waste when possible. The Dangerous Substances Directive (76/464/EEC) defines those forms of pollution as 'toxic, persistent and which bioaccumulate ... and which have a deleterious affect upon the environment' (*ibid.*).

The EPA's licensing activities cover a range of areas such as the granting of Integrated Pollution Prevention Control (IPPC) Licences, the granting of Waste Licences, overseeing emissions trading, the licensing of the use of Genetically Modified Organisms (GMOs) and Volatile Organic Compounds (VOCs) (http://www.epa.ie).

Integrated Pollution Prevention and Control (IPPC) Licensing

The original Integrated Pollution Control (IPC) licences which were first introduced in Ireland in 1994 have been replaced by the IPPC (IPPC Directives 96/6/EEC) licensed industries which have a potential to pollute. The IPPC

Directive became law in Ireland through the introduction of the Protection of the Environment (POE) Act 2003 and this Act contributes the framework for environmental regulation along with the EPA Acts of 1992 and 2003 (http://www.epa.ie). The EPA Act of 2003 introduced a licensing system which has many important requirements:

- A change in the requirement for technology which is considered as using the Best Available Technology not Entailing Excessive costs (BATNEEC) to that which was Best Available Techniques (BAT).
- Increased emphasis on energy efficiency.
- A greater emphasis on pollution prevention in the licensing system and on waste minimisation at the source.
- Provision to request an oral hearing on the granting of a license by objectors.
- Power to seek a High Court order when license agreements are being contravened.
- An open and transparent IPPC system where the public have access to application material. Objections can be made within 28 days of notification of any determination on proposals by the Agency and submissions are available for public attention (*ibid.*).

Waste Licensing

The EPA's remit also covers the licensing of the disposal and recovery, landfilling and transfer of waste in Ireland. The EPA must be satisfied that any activity involving waste will not cause environmental pollution (*ibid.*). The provisions for the licensing of waste-related activity stems from the 1992 EPA Act and the 1996 Waste Management Act. The Minister for the Environment updated the Protection of the Environment Act in 2004 which included the introduction of the following:

- Applications for Waste Licenses are subject to the Waste Management (Licensing) Regulations (2004).
- Waste Licenses to be issued under Best Available Technique (BAT) guidelines, replacing BATNEEC.
- New mechanisms to fast track license services of a minor nature.
- Energy efficiency to be considered in relation to waste licensing.
- Open and transparent waste licensing system which allows for public access to documentation; provision for written submissions or objections to applications within 28 days of proposed decision (*ibid.*).

Emissions Trading

The EPA is a participant in the European Union Emissions Trading Scheme (EU ETS) which was introduced in 2005. The EU ETS was devised as part of efforts to reduce emissions of carbon dioxide and greenhouse gases in order to reduce the threats of climate change and ozone depletion. The EU ETS provides EU Member States with a framework with which to address their commitments under the Kyoto Protocol. Under the EU ETS Member States establish an emission cap on industrial plants or installations. Industries then apply for emissions allowances under the Scheme. The Kyoto Protocol was agreed by all member states of the European Union and by the EU itself along with other global governments with the aim of reducing greenhouse gas emissions between 2008 and 2012 and beyond that date (http://www.epa.ie). The consequences of greenhouse gas emissions include the following:

- Geographic affects on species of animals, including extinction.
- Changes to rainfall patterns and subsequent supplies of fresh water.
- Increased incidents of extreme weather, floods and draughts (*ibid.*).

Licensing Genetically Modified Organism (GMOs)

Genetically Modified Organisms are the bacteria, viruses, fungi, plant and animal cells capable of replication or transfer of genetic material in an unnatural or scientifically assisted manner. The EPA oversees the implementation of GMP regulation, use, release and movement (*ibid.*). The EPA has begun to seek out land for the experimental and controlled use of GMOs at selected sites in Ireland, including a five-year experiment in County Meath.

REGULATORY ENFORCEMENT

Office of Environmental Enforcement (OEE)

The office of Environmental Enforcement (OEE) aims to improve levels of compliance with environmental regulation and legislation. The OEE is responsible for the enforcement of EPA licenses and oversees the environmental protection of the local authorities as well as other bodies including the Health Authorities, An Garda Siochana, the Criminal Assets Bureau, the Revenue Commissioners and the Office of Corporate Enforcement (http://www.epa.ie).

EPA inspectors are also involved in the enforcing of licenses issued under the Waste Management Act and the Environmental Protection Agency Act. These inspectors also monitor emissions trading and oversee the introduction of Environmental Management Systems (EMS) at industrial facilities in order to prevent pollution at source (*ibid.*).

The supervision by OEE of local authorities is undertaken through the assessment of information, audits and performances of the authorities, by directing local authorities to respond to issues within a set timeframe and through issuing sundry directives to the authorities in cases of pollution. The OEE investigates complaints of environmental laxity or regulatory failure made against the local authorities. The OEE operates in conjunction with an Enforcement Network which incorporates Government Departments, State agencies and local authorities in a comprehensive framework which assists enforcement and Best Practice. A similar European Network for EU Member States also exists (*ibid.*). In addition, the EPA is a participant in the European Network for the Implementation and Enforcement of Environmental Law (IMPEL). This network of European regulators cooperate on implementing environmental legislation across the EU Member and Candidate States by sharing data and expertise, standardising influence taken approaches and developing Best Practice (http://www.epa.ie).

The EPA Protection Act

The EPA was established by the EPA Act of 1992. The EPA emerged in response to the first phase of environmental populism in Ireland which was characterised by local opposition to multinational plants that were perceived to pose risks to the communities in which they were sited as well as to the environment (Leonard, 2005, 2006). The EPA Act was an attempt to provide an integrated model for environmental regulation based on the US or Scandinavian models which were researched as part of attempts to develop a framework for the Agency. The introduction of the EPA Act came in response to a series of issues including concerns about the absence of relevant expertise in the local authorities which had previously been charged with environmental regulation; the concerns of the public about this failure to enforce regulations adequately; the problems encountered by industry in obtaining relevant approval for developments; the need for increased regulation of local authority performance in relation to the environment and the need for clarity encountered by industry in obtaining relevant approval for developments; the need for increased regulation of local authority performance in relation to the environment and the need for clarity and fairness in the pursuit of environmental protection and regulation (Scannell, 1995, p. 513).

The EPA's constitution obliges the Agency to retain an overview of the performances of the public authorities; maintain a high standard of environmental protection; promote sustainable development and to promote precautions in relation to the potentially harmful affects of emissions; to regulate according to the polluter pays principle (*ibid.*). The EPA's authority extends to control of public authorities in relation to previously held exemptions from Planning Permission and pollution controls. Together with the Department of the Environment the EPA provides advice and a regulatory framework for the public authorities establishing codes of practice and other guidelines designed to create a standardisation in environmental regulatory enforcement.

The EPA and Political Culture in Ireland

Without doubt, the creation of the EPA in 1993 was an attempt to address the myriad 'bottom up' populist interests which were opposed to industrial plants or communities or increased regulation at a time when the 'top down' external pressure being exerted by the European Commission by way of environmental directives was increasing. The State's response was to introduce the EPA Act in 1992 creating an Agency that came to be criticised for its pro-industry leanings (Taylor, 2001) but one which, nonetheless, has developed into a facilitation of improved regulatory performance in addition to a significant contribution towards environmental research and education. However, the overlapping personal and direction of some appointees to the EPA Board has been the subject of controversy from time to time while an effective dialogue between the Agency and Environmental activists remains elusive. However, where the Agency has been criticised for having a pro-business and development perspective throughout the boom years of the Celtic Tiger, a sustained effort to deal with issues created by the post-consumption waste management crisis has also led to criticism by populist community campaigns of incinerators or landfills (Leonard, 2005, 2006).

SUMMARY

We have covered the background to the formation and subsequent performance of the Irish EPA and highlighted the fact that environmental protests have increased since the EPA's inception in 1993 despite the

Agency's remit of dealing with the sources of the pollution that cause many of these protests. The reason for ongoing concerns about the performance of the EPA is outlined as in some way due to the State's attempt to position the Agency as an all persuasive answer to the external pressures it was facing in regard to the political structures surrounding environmental policy formulation and implementation.

While the State had to create the policy frameworks necessary to allow EU environmental directives to be implemented it was also concerned with the extent of community resistance to local sitings of toxic plants and industries. However, in spite of these prominent concerns the EPA was ultimately founded in the context of the Irish State's emphasis on creating economic growth and competitiveness through its neo-corporatist partnership model whereby the attraction of multinational investment superseded any real concern for environmental degradation.

In effect, the establishment of the EPA answered the State's need to address the perception that industries in Ireland had been 'left to police themselves' in the area of pollution controls (*Irish Times*, 5 September 1989 from Taylor & Horan 1999, p. 16). In so doing the State would not compromise the operations of the multinationals or address what Greenpeace had described in 1993, the year the EPA was founded, as the 'ambivalence towards the concept of environmental protection' (Allen, 2004, p. 13) which existed in Irish Politics.

REFERENCES

Adshead, M., & Millar, M. (Eds.). (2003). *Public Administration and Public Policy in Ireland: Theory and Methods*. London: Routledge.

Allen, R. (1992). *Waste not, want not*. London: Earthscan.

Allen, R., & Jones, T. (1990). *Guests of the Nation: The People of Ireland Versus the Multinationals*. London: Earthscan.

Allen, K. (2000). *The Celtic Tiger: The myth of social partnership in Ireland*. Manchester: Manchester University Press.

Allen, R. (2004). *No global: The people of Ireland versus the multinationals*. London: Pluto.

Allen, R., & Jones, T. (1990). *Guests of the nation: The people of Ireland versus the multinationals*. London: Earthscan.

Baker, S. (1990). Irish environment. In: W. Rudig (Ed.), *Green politics 1*. Edinburgh: Edinburgh University Press.

Bax, M. (1976). *Harpstrings and confessions: Machine style politics in the Irish Republic*. Assen: Van Gorcum.

Beck, U. (1992). *Risk Society: Towards a New Modernity*. London: Sage.

Beck, U. (1995). *Ecological Politics in an Age of Risk*. Cambridge: Polity Press.

Beck, U. (1996). *Risk society: Towards a new modernity*. London: Sage.

Bilton, T., Bonnett, K., Jones, P., Lawson, T., Skinner, D., Stanworth, M., & Webster, A. (2002). *Introduction to sociology* (Vol. 4). London: Palgrave.

Blowers, A. (1987). Transition or transformation? Environmental policy under Thatcher. *Public Administration*, *65*(3), 277–294.

Breen, R., Hannon, D., Rottman, D., & Whelan, C. (1990). *Understanding contemporary Ireland*. London: MacMillan.

Carmin, J. (2003). Resources, opportunities and local environmental action in the democratic transition and early consolidation periods in the Czech Republic: *Environmental Politics*, *12*(3), 42–64.

Castells, M. (2001). *The Internet Galaxy: Reflections on the Internet, Business and Society*. London: Oxford University Press.

Cassidy, L., & Roche, B. (2011). Cork incinerator plan rejected. *Irish Times*, June, 2011.

Collins, K., & Earnshaw, D. (1992). The implementation and enforcement of European community environment legislation. *Environmental Politics*, *1*(4), 213–249.

Christoff, P. (1996). Ecological modernisation, ecological modernities. *Environmental Politics*, *5*(3), 476–500.

Coakley, J., & Gallagher, M. (Eds.). (1999). *Politics in the Republic of Ireland* (3rd ed.). London: PSAI Press.

Collins, N., & O'Shea, M. (2003). Clientelism: Facilitating rights and favours. In: M. Adshead & M. Millar (Eds.), *Public Administration and Public Policy in Ireland: Theory and Methods*. London: Routledge.

Cox, L. (1997). The future of the Irish environmental movement. *1st Irish Environmental Futures Conference*, NUI Galway, Galway.

Cox, L. (1999). Structure, routine and transformation: Movements from below at the end of the century. In: C. Barker & M. Tyldersley (Eds.), *Fifth international conference on alternative futures and popular protest*. Manchester: Manchester University Press.

Cox, L. (Ed.) (2001). *Globalisation from Below? "Ordinary People", Movements and Intellectuals*. Dublin: RIAN.

Coyle, C. (1994). Administrative capacity and the implementation of EU environmental policy in Ireland. In: S. Baker, K. Milton & S. Yearly (Eds.), *Protecting the periphery: Environmental policy in peripheral regions of the EU*. London: Frank Cass.

Crossley, N. (2002). *Making sense of social movements*. Philadelphia, PA: Open University Press.

Cunnane, K. (2005). Galway City Community Forum Newsletter – Tribe Vibes. Galway: Galway City Forum.

Daugbjerg, C., & Svendsen, G. T. (2003). Designing green taxes in a political context: From optimal to feasible environmental regulation. *Environmental Politics*, *12*(4).

Dearing, J. W., & Rogers, E. M. (1996). *Agenda-Setting*. Thousand Oaks, CA: Sage.

Della Porta, D., & Diani, M. (1999). *Social movements: An introduction*. London: Blackwell.

Della Porta, D., & Diani, M. (2006[1999]). *Social Movements* (2nd revised and expanded edition). Oxford/Malden, MA: Blackwell.

Denison, R. A., & Ruston, J. (1990). *Recycling and Incineration: Evaluating the Choices*. Washington, D.C.: Island Press.

Doherty, B., & DeGeus, M. (1996). *Democracy and green political thought*. London: Routledge.

Dryzek, J. *The politics of the earth*. London: Earthscan.

Dryzek, J., Downes, D., Scholosberg, D., Herens, H., & Hurold, C. (2002). Environmental transformation of the State: USA, Norway, Germany and the UK. *Political Studies*, *50*, 659–682.

Eckersley, R. (1992). *Environmentalism and political theory: Towards an ecocentric approach*. London: UCL Press.

Eisinger, P. (1973). The conditions of protest behaviour in American cities. *America Political Science Review*, *67*, 11–28.

Environmental Protection Agency. (1996). *Waste Management Act*. Dublin: EPA.

Environmental Protection Agency. (2002). *Millennium Report*. Dublin: EPA.

EPA National Waste Database (NWD) Report. Dublin: EPA.

Etzioni, A. (1968). *The active society*. New York, NY: Free Press.

Fagan, G. H. (2003). Sociological reflections on governing waste. *Irish Journal of Sociology*, *12*(1).

Fagan, O'Hearn, D., McCann, G., Murray, M., & Cross, A. (2001). *2001 Waste management strategy: Border perspective*. Maynooth: National Institute for Regional and Spatial Analysis.

Feagin, J., Orum, A., & Sjoberg, G. (Eds.). (1991). *A case study for case study*. Chapel Hill, NC: University of North Caroline Press.

Friedman, D., & McAdam, D. (1992). In: C. Mueller & A. Morris (Eds.), *Frontiers of social movement theory*. New Haven, CN: Yale University.

Fry, J. (1999). Can EIA deliver sustainable development? *Irish Planning and Environmental Law Journal*, *6*(1), 17–21.

Gallagher, M., March, M., & Mitchell, P. (2003). *How Ireland voted 2002*. London: Palgrave.

Galway For a Safe Environment. (1999). *A burning issue*. Galway: GSE.

Gamson, W. (1990). *The strategy of social protest*. Belmont, CA: Wadsworth.

Gamson, W. A. (1975). *The Strategy of Social Protest*. Homewood, IL: Dorsey.

Gamson, W. A., & Meyer, D. S. (1996). Framing political opportunity. In: D. McAdam, J. D. McCarthy & M. N. Zald (Eds.), *Comparative Perspectives on Social movements: Political Opportunities, Mobilizing Structures, and Cultural framings* (pp. 273–290). Cambridge: Cambridge University Press.

Garavan, M. (2004). *The patterns of Irish environmentalism*. Ph.D. thesis, NUI, Galway.

Goffman, E. (1974). *Frame analysis*. Cambridge, MA: Harvard University Press.

Government of Ireland. (1998). *Changing our ways,* Policy Statement. Department of Environment and Local Government, Minister Noel Dempsey, Dublin.

Green Party. (2002). Election manifesto. *Meeting the challenges*. Dublin.

Grove-White, R. (1993). Environmentalism: A new moral discourse for technological society? In: K. Milton (Ed.), *Environmentalism: The View from Anthropology* (pp. 18–30). London: Routledge.

Hajer, M. (1995). *The politics of environmental discourse. Ecological modernisation and the policy process*. Oxford: Clarendon.

Hamilton, D. (1992). *Inward investment in Northern Ireland*. Report no. 99. Belfast: Northern Ireland Economic Council.

Held, D. (1987). *Models of democracy*. London: Polity Press.

Hirst, P. (1993). *Associational democracy*. London: Routledge.

IDEA. (1999). *Irish Doctor's Environmental Association Information Pamphlet: 'Incinerators and Their Health Effects'*. Dublin: IDEA.

Inglehart, R. (1977). The silent generation in Europe: Inter-generational change in post-industrial societies. *American Political Science Review*, *65*, 991–1017.

Janicke, M. (1999). Democracy as a condition for policy success. From democracy and the environmental problems and prospects. In: Lafferty & Meadowcroft (Eds.), Cheltenham: Edward Elgar.

Jessop, B. (1990). *State theory: Putting the capitalist state in its place*. Cambridge: Polity.

Jordan, G., & Richardson, J. (1982). The British policy style or the logic of negotiation? In: J. Richardson (Ed.), *Policy Styles in Western Europe* (pp. 80–110). London: George Allen and Unwin.

Klandermans, B. (Ed.). (1989). *Organizing for change: Social movement organizations in Europe and the United States*. International Social Movement Research. Greenwich, CT: JAI Press.

Klandermans, B. (1988). The formation and mobilization of consensus. In: B. Klandermans, H. Kriesi & S. Tarrow (Eds.), *From structure to action: Comparing social movements across cultures*. Greenwich, CT: JAI Press.

Klein, N. (2000). *No Logo*. London: Flamingo.

Kriesi, H. (1995). The political opportunity structure of new social movements: Its impact on their mobilisation. In: J. Jenkins & B. Klandermans (Eds.), *The politics of social protest*. London: UCL.

Kriesi, H., & Tarrow, S. (Eds.). (1990). *From Structure to Action: Social Movement Participation Across Cultures* (pp. 219–244). Greenwich, CT: JAI Press.

Lafferty, W., & Hovden, E. (2003). Environmental policy integration: Towards an analytical framework. *Environmental Politics*, *12*(3).

Laurence, D. (2000). Swallows and fishes: The definition of waste in the WMA 1996. *Irish Planning and Environmental Law Journal*, *7*(43).

Laver, M. (Ed.) (2001). *Estimating the policy position of political actors.* London: Routledge.

Leonard, H. J. (1988). *Pollution and struggle for the world product.* Cambridge: University Press.

Leonard, L. (1999). *Ecolaiocht na hEireann: Irish environmental policy.* M. Phil. Masters thesis, NUI, Galway.

Leonard, L. (2005). Politics Inflamed: Galway for a Safe Environment (GSE) and the Campaign Against Incineration in Ireland, Ecopolitics Series Volume 1. Galway: Greenhouse Press in association with Choice Publishing, Drogheda.

Leonard, L. (2006). *Green nation: The Irish environmental movement in Ireland from Carnsore Point to the Rossport 5.* Galway: Greenhouse Press.

Leonard, L. (2008). *The environmental movement in Ireland.* Dordrecht: Springer.

Leonard, L. (2010). The Galway water crisis of 2007. In: M. Corcoran & P. Share (Eds.), *Sociological chronicles 5.* Dublin: Liffey Press.

Leonard, L, Doran, P., & Fagan, H. (2010). A burning issue? Governance and anti-incinerator campaigns in Ireland, North and South. *Environmental Politics, 18*(6); In C. Rootes & L. Leonard (Eds.), *Environmental movements and waste infrastructure.* London: Taylor/ Francis.

Lowe, P., & Ward, S. (Eds.). (1998). *British Environmental Policy and Europe: Politics and Policy in Transition.* London: Routledge.

Lyotard, J. F. (1984). *The postmodern condition: A report on knowledge.* Manchester: Manchester University Press.

Mair, P. (2001). Searching for the positions of political actors: A review of approaches and a critical evaluation of expert surveys. In: M. Laver (Ed.), *Establishing the positions of political actors.* London: Routledge.

Majone, G. (1997). The Agency Model: The Growth of Regulation and the European Union. *Journal of European Public Policy, 4,* 2.

Martell, L. (1994). *Nature and society: An introduction.* Cambridge: Polity.

McAdam, D. (1982). *Political Process and the Development of Black Insurgency, 1930–1970.* Chicago: The University of Chicago Press.

McAdam, D., McCarthy, J., & Zald, M. (Eds.). (1996). *Comparative perspectives on social movements, political opportunities, mobilising structures and cultural framing.* Cambridge, NY: Cambridge University Press.

McCarthy, J., & Zald, M. (1977). Resource mobilisation and social movements. *American Journal of Sociology, 82.*

McDonagh, J., Varley, T., & Shortall, S. (2009). *A living countryside – The politics of sustainable development in rural Ireland.* Aldershot: Ashgate.

Melucci, A. (1996). *Challenging codes: Collective Action in the Information Age.* Cambridge: Cambridge University Press.

Melucci, A. (1988). *Nomads of the Present: Social Movements and Individual Needs in Contemporary Society.* London: Temple University Press.

Michael Skou, A., & Liefferink, D. (Eds.). (1997). *European Environmental Policy: The Pioneers.* Manchester: Manchester University Press.

Mullally, G. (1997). Treading softly on the political system: The Irish Greens in the 1997 general election. *Environmental Politics, 6*(4).

Mullally, G. (1999). *Second Regional Conference on Social Movements and Change: Ireland, Social Movements in Transition: Moving Towards the Millennium.* Relocating Protest: The Institutionalisation of Organised Environmentalism in Ireland, Department of Sociology, University College.

O'Dowd, L., & Wilson (Eds.). (1996). *Borders, nations and states: Frontiers and sovereignty in the New Europe*. Aldershot: Avebury.

Offe, C. (1987). Changing boundaries of institutional politics: Social movements since the 1960s. In: C. S. Maier (Ed.), *Changing Boundaries of the Political*. Cambridge: Cambridge University Press.

Office of Technology Assessment, US Congress. (1989). Facing America's trash: What's next for municipal solid waste 219.

O'Hearn, D. (1999). *Inside the Celtic Tiger: The Irish economy and the Asian model*. London: Pluto.

O'Riordan, T. (1981). *Environmentalism*. London: Pion, Zedn.

O'Riordan, T. (1989). The challenge for environmentalism. In: R. Peet, & N. Thrifts (Eds.), *New Models in Geography 48* (pp. 33–46).

O'Sullivan, J. (2000). An approach to resolving the current waste crisis. *Irish Environmental and Planning Law Journal*, *8*(1), 20–25.

Peace, A. (1993). Environmental protest, bureaucratic disclosure: The politics of disclosure in rural Ireland. In: D. Milton (Ed.), *Environmentalism: The view from anthropology*. London: Routledge.

Piven, F. F., & Cloward, R. A. (1979). *Poor People's Movements: Why They Succeed, How They Fail*. New York, NY: Vintage.

Progressive Democrats. (2002). Election manifesto. *Value for vote: Modernising waste management*. Dublin.

Rabe, B. (1994). *Beyond NIMBY: Hazardous waste siting in Canada and the United States*. Washington, DC: Brookings.

Rawcliffe, P. (1998). *Environmental pressure groups in transition*. Manchester: MUP.

Roberts, J. (2004). *Environmental policy*. London: Routledge.

Roche, W. K., & Cradden, T. (2003). Neo-corporatism and social partnership. In M. Ashead & M. Millar (Eds.), *Public administration and public policy in Ireland: Theory and methods*. London: Routledge.

Rogers, E. M., & Dearing, J. W. (1988). Agenda-setting research: Where has it been? Where is it going? In: J. A. Anderson (Ed.), *Communication Yearbook 11* (pp. 555–594). Newbury Park, CA: Sage.

Rootes, C. (1997a). Environmental movements and green parties in Western and Eastern Europe. In: M. Redclift & G. Woodgate (Eds.), *International handbook of environmental sociology*. Cheltenham: Edward Elgar.

Rootes, C. (1997b). Shaping collective action: Structure, contingency and knowledge. In: R. Edmondson (Ed.), *The political context of collective action: Power, argumentation and democracy*. London: Routledge.

Rootes, C. (1999). *Environmental Movements: Local, National and Global*. London: Frank Cass.

Rootes, C., & Leonard, L. (2010). *Environmental movements and waste infrastructure*. London: Routledge.

Sbragia, A. (1996). Environmental policy: The "push-pull" of policy-making. In: H. Wallace & W. Wallace (Eds.), *Policy-Making in the European Union* (pp. 235–255). Oxford: Oxford University Press.

Scannell, Y. (1995). *Environmental and Planning Law in Ireland*. Dublin: Roundhall Press.

Scott, A. (1990). *Ideology and the new social movements*. London: Unwin Hyman.

Scruggs, L. (1999). Institutions and environmental performance in 17 western democracies. *British Journal of Political Studies*, *29*, 1–31.

Scharpf, F. W. (1984). Economic and institutional constraint of full employment strategies: Sweden, Austria and West Germany 1973–1982. *Order and conflict in contemporary capitalism*. Oxford: Oxford University Press.

Smyth, J. J., & Boylan, T. A. (1991). Industrialisation and the contribution of multinational companies to economic development in County Mayo. In: T. Varley, T. A. Boylan & M. P. Cuddy (Eds.), *Rural crisis: Perspectives on Irish Rural Development* (pp. 156–171). Galway: Centre for Development Studies, University College Galway.

Snow, D., & Benford, R. (1988). Ideology, frame resonance and participant mobilisation. In: B. Klandermans, H. Kriesi, & S. Tarrow (Eds.), *Structures to action: Comparing social research across cultures* (pp. 197–218). International Social Research, Vol. 1. Greenwich Common: JAI Press.

Snow, D., & Benford, R. (1992) Master frames and cycles of protest. In: A. D. Morris & C. M. Mueller (Eds.), *Frontiers in social movement theory*. New Haven, Conn: Yale University Press.

Snow, D., Burke, R., Warden, S., & Benford, R. (1986). Frame alignment processes, micromobilisation and movement participation. *American Sociological Review*, *51*, 464–481.

Stake, R. (1995). *The art of case research: Design and methods*. Newbury Park, CA: Sage.

Stammers, N. (1999). Social movements and the social construction of human rights. *Human Rights Quarterly*, *21*(4).

Szasz, A. (1994). *Eco-populism: Toxic waste and the movement for environmental justice*. London: UCL Press.

Tarrow, S. (1984; 1988). *Power in Movement: Social Movements and Contentious Politics* (2nd ed.). Cambridge studies in comparative politics. Cambridge: Cambridge University Press.

Tarrow, S. (1988). National politics and collective action: Recent theory and research in Western Europe and the United States. *Annual Review of Sociology*, *14*.

Tarrow, S. (1994). *Power in movement: Social movements, collective action and politics*. New York, NY: Cambridge University Press.

Tarrow, S. (1998). *Power in movement: Social movements and contentious politics* (2nd ed.). Cambridge: CUP.

Taylor, G. (1998). Me thinks thou do'st protest too much: Oral hearings, public registers and environmental democracy in Ireland. *Irish Planning and Environmental Law Journal*, *5*(4), 143–155.

Taylor, G. (1998). Conserving the Emerald Tiger: The politics of environmental regulation in Ireland. *Environmental Politics*, *7*(4), 53–74.

Taylor, G. (2001). Looking back (but not in anger): Policy transfer and the role of institutional memory in the construction of the Irish EPA. *Irish Studies in International Affairs*, *13*, 187–201. Royal Irish Academy.

Taylor, G. (2001). *Conserving the Emerald Tiger: The Politics of Environmental Regulation in Ireland*. Arlen Academic Press.

Taylor, G. (2002). Hailing with an invisible hand: A 'Cosy' political dispute amid the rise of neo-liberalism in modern Ireland. *Government and Opposition*, *37*(4), 501–523.

Taylor, G., & Flynn, B. (2003). It's green, but is it of a light enough hue? Past performance, present success and the future of the Irish Greens. *Environmental Politics*, *12*(1).

Taylor, G., & Murphy, C. (2002). Environmental policy. In: Taylor (Ed.), *Issues in Irish Public Policy*. Irish Academic Press.

Taylor, G., & Horan, A. (1999). *From cats, dogs, parks and playgrounds to IPC licensing: Policy learning and the evolution of environmental policy in Ireland*. Galway: NUIG.

Taylor, V., & Whitter, N. (1992). Master frames and cycles of protest. In: A. Morris & M. C. Mueller (Eds.).

Tilly, C. (1977). *From Mobilization to Revolution*. Reading, MA: Addison Wesley.

Tilly, C. (1978). *From mobilisation to revolution*. New York, NY: McGraw-Hill.

Tovey, H. (1992). Environmentalism in Ireland. In P. Clancy, M. Kelly, J. Wiatr, & R. Zoltaniecki (Eds.), *Ireland and Poland: Comparative perspectives*. Dublin: UCD.

Tovey, H. (1999). Rural poverty: A political economy perspective. In: D. G. Pringle, J. Walsh & M. Hennessy (Eds.), *Poor People Poor Places*. Dublin: Oak Tree.

Tovey, H., & Share, P. (2000). *Sociology of Ireland*. Dublin: Gill and MacMillan.

Turner, R. H., & Killian, L. M. (1987). *Collective Behavior*. Englewood Cliffs, NJ: Prentice-Hall.

Van der Straaten, J. (1992). The Dutch National Environmental Policy Plan: To choose or to lose. *Environmental Politics*, *1*(I), 4.5–4.7.

Van der Heijden, H. A. (1997). Political opportunity structure and the institutionalisation of the environmental movement: *Environmental Politics*, *6*(4), 25–50.

Varley, T., Boylan, T., & Cuddy, M. (Eds.). (1991). *Rural crisis: Perspectives on Irish Rural Development* (pp. 156–171). Galway: Centre for Development Studies, University College Galway.

Walsh, E., Warland, R., & Smith, D. (1997). *Don't burn it here: Grassroots challenges to trash incinerators*. University Park, PA: Pennsylvania University Press.

Weale, A. (1992). *The new politics of pollution*. Manchester: Manchester University Press.

Wexford IFA (1999, December). *Proceedings of the Agri-Food Millennium Conference*, 7 December, Irish Farmer's Association, Wexford.

Whiteman, D. (1990). The progress and potential of the Green Party in Ireland. *Irish Political Studies*, *5*.

Wilson, J. (1973). *Introduction to social movements*. New York, NY: Basic Books.

World Commission on Environment and Development. (1987). Our common future, Oxford: Oxford University Press.

Yearley, S. (1995). The shaping of the environmental movement in Ireland. In: P. Clancey, S. Drudy, K. Lynch & L. O'Dowd (Eds.), *Irish Society: A sociological perspective*. Dublin: Institute of Public Administration.

Yearley, S. (1996). *Sociology, Environmentalism, Globalization*. London: Sage.

Yin, R. (1994). *Case study research: Design and methods*. Thousand Oaks, CA: Sage.

Zald, M. N., & McCarthy, J. D. (1987). *Social Movements in an Organizational Society*. New Brunswick, NJ: Transaction Books.

Newspaper and Media Sources:

Connacht Sentinel
Connacht Tribune
Cork Examiner
Galway Advertiser
Galway Bay F.M.
Galway Independent
Irish Independent
Irish Times
Japan Times
Mainichhi Dáily News
RTÉ Six One News
Tuam Herald
Washington Times
GSE website: http://go.to/gse
No Incineration Alliance (2002) *Appeal Against the Duleek Incinerator Site*. Meath: No Incineration Alliance.

Interviews with Relevant Participants

(i) Conchúr Ó Brádaigh
(ii) Michael D. Higgins
(iii) Joe Gavin
(iv) Professor Paul Connett
(v) Professor Emer Colleran
(vi) Aine Suttle

AUTHOR INDEX

SUBJECT INDEX